Student Solutions Manual
for Urone's
COLLEGE PHYSICS

SECOND EDITION

REGINA L. NEIMAN
S.H. Lee KnowledgeStream

STEVEN YOUNT
U.S. Patent Office

PAUL PETER URONE
California State University, Sacramento

BROOKS/COLE

THOMSON LEARNING

Australia • Canada • Mexico • Singapore • Spain • United Kingdom • United States

BROOKS/COLE

THOMSON LEARNING

Assistant Editor: *Samuel Subity*
Marketing Assistant: *Stephanie Rogulski*
Production Coordinator: *Dorothy Bell*
Cover Design: *Roy R. Neuhaus*

Cover Photo: *Jim Cummins/FPG International*
Print Buyer: *Micky Lawler*
Printing and Binding: *Globus Printing*

For more information about this or any other Brooks/Cole product, contact:
BROOKS/COLE
511 Forest Lodge Road
Pacific Grove, CA 93950 USA
www.brookscole.com
1-800-423-0563 (Thomson Learning Academic Resource Center)

For permission to use material from this work, contact us by
Web: www.thomsonrights.com
fax: 1-800-730-2215
phone: 1-800-730-2214

Printed in the United States of America

10 9 8 7 6 5 4 3 2

ISBN 0-534-37690-8

CONTENTS

PREFACE

This manual presents answers to one third of the conceptual questions and solutions to one third of the odd-numbered end-of-chapter problems in Urone's *College Physics, Second Edition*. The purpose of this manual and of the end-of-chapter problems is to build problem-solving skills that are one fundamental aspect of understanding and applying the principles of physics. The following paragraphs will help you to use these solutions to build your understanding of physics.

Use Your Text to Your Advantage

Your text contains many features that will help you not only to solve problems, but also to understand their concepts. Before turning to a problem solution you should use these features in your text to your advantage. The worst thing that you can do with these solutions is to copy them directly without thinking about the problem-solving process and concepts involved.

Problem-Solving Strategies. The *problem-solving strategies* in the text present techniques that will aid you in solving problems. You can find these strategies listed in the *index* of the text. The problem-solving strategies ask you to go beyond what is presented in the solutions in this manual. The purpose of following these strategies is to build your problem-solving skills so that you can solve homework problems, be able to apply physics outside your course work, and succeed at exams.

Worked Examples. Numerous *worked examples* appear as part of the regular text. Some also appear in the end-of-chapter material to aid in the understanding of integrated concepts problems and unreasonable results problems. Follow the worked examples carefully, since they relate to specific principles and have a *discussion* section that elaborates on the meaning of the example.

Chapter Summaries. The *summary* at the end of each chapter contains all important definitions and equations presented in the chapter. Consulting the summary can help build an understanding of the basics of the chapter.

Index. The *index* at the end of the text is user friendly and highly detailed. In it you can find topics related to the problem you are attempting to solve as well as a wealth of information pertinent to all subjects. It may be particularly useful to consult the index when solving integrated concepts problems. As mentioned above, you can also find relevant problem-solving strategies in the index.

Glossary. The *glossary* at the end of the text gives definitions of every symbol used in the text. You will also find the section of the text and the page on which the symbol is defined. This is particularly useful when the same symbol stands for several different physical quantities, such as T for period, temperature, and tension.

vi   College Physics, Second Edition

Chapter Material. Read the chapter material and notes taken in your class before attempting the end-of-chapter problems. A brief review of that material is very useful when you attempt the problems. You will find that when you go over the chapter material a second or third time, you may pick up on items that at first escaped you. But most of all, in the chapter material you will find the discussions, definitions, applications, and illustrations that tell the story of the physics involved in the problems.

A Wish for Your Success

I wish you true success in your study of physics. Solving homework problems successfully is a major step in learning the principles of physics and how to apply them. It involves active inquiry, exploration, extrapolation, and integration of ideas and concepts. It is where you understanding of physics gels. The more you succeed at solving physics problems, the more you will enjoy the wonders of the discipline.

Paul Peter Urone

INTRODUCTION ①

CONCEPTUAL QUESTIONS

1.1 A model is a mental image or analogy to objects or phenomena that we can experience directly; it can be crucial to visualizing phenomena that we cannot observe directly.

1.4 The validity of a theory is determined by observation and experiment. If a theory is valid, then it is consistent with all observations and experiments.

1.7 Classical physics is a good approximation to modern physics when objects move slowly compared to the speed of light (less than about 1% of the speed of light).

1.10 "I am 5 10, how tall are you?" Implies 5 *feet* 10 *inches*.

1.13 Accuracy is how close a measurement is to the "true" value. The uncertainty in a measurement is an estimate of the amount by which it differs from the "true" value. Therefore, the uncertainty is a measure of the accuracy of the value.

1.16 The power of a lens has the unit of diopters (D). The minimum uncertainty with which corrections are determined is ¼ diopter. The accuracy with which corrective lenses can be produced is called the tolerance, and the tolerance for eyeglasses is 1/8 diopter. The uncertainties in the prescription come about because of when determining the power of the eye, the eye muscles can compensate for some of the irregularities in the shape of the eye. The accuracy in the manufacturing of lenses is limited by the shape of the lens itself, and how large of a range over which it must be accurate. At the center of a lens, the lens is more accurately made than near the edges.

PROBLEMS

1.1 Yes, classical physics is an excellent approximation when objects move at speeds less than about 1% of the speed of light $(3.00 \times 10^8 \text{ m/s})$. So if the satellite's speed is less than 1% of the speed of light, we can accurately describe it using classical physics. Since,

$$\frac{\text{satellite speed}}{\text{speed of light}} = \frac{7500 \text{ m/s}}{3.00 \times 10^8 \text{ m/s}} = 2.50 \times 10^{-5} = \boxed{2.50 \times 10^{-3} \text{ %}} ,$$

the satellite can be described classically.

1.7 (a) Using the information from Table 1.3:

$$\frac{\text{distance to nearest galaxy}}{\text{diameter of Milky Way galaxy}} = \frac{10^{22}\ \text{m}}{10^{21}\ \text{m}} = \boxed{10}.$$

The distance to the nearest galaxy (Andromeda) is 10 times larger than the size of the Milky Way galaxy.

(b) Using the data provided in Table 1.3:

$$\frac{\text{diameter of the Milky Way galaxy}}{\text{distance from earth to the sun}} = \frac{10^{21}\ \text{m}}{10^{11}\ \text{m}} = \boxed{10^{10}}.$$

The size of the Milky Way galaxy is 10^{10} times larger than the distance from the earth to the sun.

(c) Using the information in Table 1.3:

$$\frac{\text{distance to Andromeda}}{\text{distance to the sun}} = \frac{10^{22}\ \text{m}}{10^{11}\ \text{m}} = \boxed{10^{11}}.$$

The distance to Andromeda is 10^{11} times larger than the distance from the earth to the sun.

1.13 The total mass of the known universe is $10^{53}\ \text{kg}$, and the average mass of a star is $10^{30}\ \text{kg}$, so dividing one half of the mass of the universe by the average mass of a star will give the approximate number of stars in the universe.

$$\frac{1}{2} \times \frac{10^{53}\ \text{kg / universe}}{10^{30}\ \text{kg / star}} = \boxed{5 \times 10^{22}\ \text{stars / universe}}.$$

There are approximately 5×10^{22} stars in the universe, which is approximately the same as the number of atoms as in one gram of carbon, or the number of breaths all humanity has taken in the last 2000 years!

1.19 Since 3 feet = 1 yard and 3.281 feet = 1 meter, multiply 100 yards by these conversion factors to cancel the units of yards, leaving the units of meters:

$$100\ \text{yd} = 100\ \text{yd} \times \frac{3\ \text{ft}}{1\ \text{yd}} \times \frac{1\ \text{m}}{3.281\ \text{ft}} = \boxed{91.4\ \text{m}}.$$

A football field is 91.4 m long.

1.25 (a) The average speed of the earth's orbit around the sun is calculated by dividing the distance traveled by the time it takes to go one revolution:

$$\text{average speed} = \frac{2\pi(\text{average dist of Earth to sun})}{1\ \text{year}} = \frac{2\pi\left(1.50 \times 10^8\ \text{km}\right)}{365.25\ \text{d}} \times \frac{1\ \text{d}}{24.0\ \text{h}} \times \frac{1\ \text{h}}{3600\ \text{s}} = \boxed{29.9\ \text{km / s}}.$$

The earth travels at an average speed of 29.9 km/s around the sun.

(b) To convert the average speed into units of feet/s, use the conversion factors: 1000 m = 1 km and 1 foot = 0.3048 m:

$$\text{average speed} = \frac{29.9\ \text{km}}{\text{s}} \times \frac{1000\ \text{m}}{1\ \text{km}} \times \frac{1\ \text{ft}}{0.3048\ \text{m}} = \boxed{9.80 \times 10^4\ \text{ft / s}}$$

1.31 (a) To calculate the number of beats she has in 2.0 years, we need to multiply 72.0 beats/minute by 2.0 years and use conversion factors to cancel the units of time:

$$\frac{72.0 \text{ beats}}{\text{min}} \times \frac{60.0 \text{ min}}{1.00 \text{ h}} \times \frac{24.0 \text{ h}}{1.00 \text{ d}} \times \frac{365.25 \text{ d}}{1.00 \text{ y}} \times 2.0 \text{ y} = 7.5738 \times 10^7 \text{ beats}$$

Since there are only 2 significant figures in 2.0 years, we must report the answer with 2 significant figures:

$$\boxed{7.6 \times 10^7 \text{ beats}} .$$

(b) Since we now have 3 significant figures in 2.00 years, we now report the answer with 3 significant figures:

$$\boxed{7.57 \times 10^7 \text{ beats}}$$

(c) Even though we now have 4 significant figures in 2.000 years, the 72.0 beats/minute only has 3 significant figures, so we must report the answer with 3 significant figures:

$$\boxed{7.57 \times 10^7 \text{ beats}} .$$

1.37 To calculate the heart rate, we need to divide the number of beats by the time and convert to beats per minute.

$$\frac{\text{beats}}{\text{minute}} = \frac{40 \text{ beats}}{30.0 \text{ s}} \times \frac{60.0 \text{ s}}{1.00 \text{ min}} = 80 \text{ beats / min} .$$

To calculate the uncertainty, we use the **method of adding percents**.

$$\% \text{ unc} = \frac{1 \text{ beat}}{40 \text{ beats}} \times 100 \% + \frac{0.5 \text{ s}}{30.0 \text{ s}} \times 100 \% = 2.5 \% + 1.7 \% = 4.2 \% = 4\%$$

Then, to calculate the uncertainty in beats per minute, we need to δ.

$$\delta A = \frac{\% \text{ unc}}{100 \%} \times A = \frac{4.2\%}{100\%} \times 80 \text{ beats / min} = 3.3 \text{ beats / min} = 3 \text{ beats / min}$$

Notice that while doing calculations, we keep one EXTRA digit, and round to the correct number of significant figures only at the end. So, the heart rate is $\boxed{80 \pm 3 \text{ beats / min}}$.

1.43 The area = length × width = 3.955 m × 3.050 m = 12.06 m^2 . To calculate the uncertainty in the area, we use the **method of adding percents**.

$$\% \text{ unc length} = \frac{0.005 \text{ m}}{3.955 \text{ m}} \times 100 \% = 0.13 \%$$

$$\% \text{ unc width} = \frac{0.005 \text{ m}}{3.050 \text{ m}} \times 100 \% = 0.16 \%$$

$$\% \text{ unc area} = \% \text{ unc length} + \% \text{ unc width} = 0.13 \% + 0.16 \% = 0.29 \% = 0.3 \%$$

Finally, using the % uncertainty for the area, we can calculate the uncertainty in square meters:

$$\delta \text{area} = \frac{\% \text{ unc area}}{100 \%} \times \text{area} = \frac{0.29 \%}{100 \%} \times 12.06 \text{ m}^2 = 0.035 \text{ m}^2 = 0.03 \text{ m}^2$$

So, the area is $\boxed{12.06 \pm 0.03 \text{ m}^2}$

2 KINEMATICS

CONCEPTUAL QUESTIONS

2.1   These numbers are consistent because the distance measured by reconnaissance is directly (the way the crow flies) from San Francisco to Sacramento, while the actual driving distance is along the roads. We expect the actual driving distance to be longer than the direct path found by reconnaissance.

2.4   In 1967, the atomic clock became the standard for measuring the SI unit of time, the second. A second was defined as the interval required for 9,192,631,770 vibrations of the cesium-133 atom. So the vibration of the cesium-133 atom is the change that indicates a change in time.

2.7   The instantaneous velocity is the velocity at a specific instant or the average velocity for an infinitesimal or zero time interval. Instantaneous speed is defined to be the magnitude of instantaneous velocity.

2.10   No, since the acceleration is the rate at which velocity changes, if the velocity is constant, there is no change, so the acceleration must be zero.

2.13   The acceleration is positive if it reduces the magnitude of a negative velocity. The acceleration is negative if it reduces the magnitude of a positive velocity.

2.16   The rock will have the same speed on the way down as it did on the way up. The rock is probably still more likely to dislodge the coconut on the way up, even though the speeds are the same, because on the way up it would cause compression of the stem which is holding the coconut on, whereas it would push down on the coconut at askance angle to the stem and not be quite as effective.

2.19   The maximum height is inversely proportional to the acceleration for the same takeoff speed (see Equation 2.11), so an astronaut could jump six times higher on the moon than on the earth because the gravity on the moon is 1/6 of that on the earth.

2.22   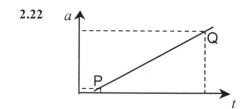

The acceleration is the slope of the velocity versus time curve, so the slope at point P is slightly larger than zero, and the slope at point Q is quite large. The slope seems to be increasing linearly, since the velocity curve appears to be parabolic.

PROBLEMS

2.1 (a) The total distance traveled is the *length* of the line from the dot to the arrow in path A, or $\boxed{7\text{ m}}$.

(b) The distance from start to finish is the magnitude of the difference between the position of the arrow and the position of the dot in path A: $|\Delta x| = |x_2 - x_1| = |7\text{ m - 0 m}| = \boxed{7\text{ m}}$

(c) The displacement is the difference between the value of the position of the arrow and the value of the position of the dot in path A: The displacement can be either positive or negative.

$$\Delta x = x_2 - x_1 = 7\text{ m} - 0\text{ m} = \boxed{+7\text{ m}}$$

2.7 (a) The average speed is defined as the total distance traveled divided by the elapsed time, so that:

$$\text{average speed} = \frac{\text{distance traveled}}{\text{time elapsed}} = 2.20 \times 10^6 \text{ m}/\text{s}.$$

If we want to find the number of revolutions per second, we need to know how far the electron travels in one revolution.

$$\frac{\text{distance traveled}}{\text{revolution}} = \frac{2\pi r}{1\text{ rev}} = \frac{2\pi\left[(0.5)\left(1.06 \times 10^{-10}\text{ m}\right)\right]}{1\text{ rev}} = \frac{3.33 \times 10^{-10}\text{ m}}{1\text{ rev}}.$$

So to calculate the number of revolutions per second, we need to divide the average speed by the distance traveled per revolution, thus canceling the units of meters:

$$\frac{\text{rev}}{\text{s}} = \frac{\text{average speed}}{\text{distance / revolution}} = \frac{2.20 \times 10^6 \text{ m}/\text{s}}{3.33 \times 10^{-10}\text{ m}/\text{rev}} = \boxed{6.61 \times 10^{15}\text{ rev}/\text{s}}.$$

(b) The velocity is defined to be the displacement divided by the time of travel, so since there is no NET displacement during any one revolution, $\boxed{v = 0\text{ m/s}}$.

2.13 The average velocity for the first segment, is the distance traveled downfield (the positive direction) divided by the time he traveled:

$$\overline{v}_1 = \frac{\text{displacement}}{\text{time}} = \frac{+15.0\text{ m}}{2.50\text{ s}} = \boxed{+6.00\text{ m/s}}\text{ (forward)}.$$

The average velocity for the second segment, is the distance traveled (this time in the negative direction because he is traveling backward) divided by the time he traveled:

$$\overline{v}_2 = \frac{-3.00\text{ m}}{1.75\text{ s}} = \boxed{-1.71\text{ m/s}}\text{ (backward)}.$$

Finally, The average velocity for the third segment, is the distance traveled (positive again because he is again traveling downfield) divided by the time he traveled:

$$\overline{v}_3 = \frac{+21.0\text{ m}}{5.20\text{ s}} = \boxed{+4.04\text{ m/s}}\text{ (forward)}.$$

Finally, to calculate the average velocity for the entire motion, we add the displacements from each of the three segments (remembering the sign of the numbers), and divide by the total time for the motion:

$$\overline{v}_{\text{total}} = \frac{15.0\text{ m} - 3.00\text{ m} + 21.0\text{ m}}{2.50\text{ s} + 1.75\text{ s} + 5.20\text{ s}} = \boxed{+3.49\text{ m/s}}.$$

Notice that the average velocity for the entire motion is NOT just the addition of the average velocities for the segments!

2.19 Given: $a = -2.10 \times 10^4 \text{ m/s}^2$; $t = 1.85 \text{ ms} = 1.85 \times 10^{-3}$ s; $v = 0$ m/s, find v_0. We use Equation 2.9 because it involves ONLY terms we know and terms we want to know: $v = v_0 + at$. Solving for our unknown gives:

$$v_0 = v - at = 0 \text{ m/s} - \left(-2.10 \times 10^4 \text{ m/s}^2\right) \times \left(1.85 \times 10^{-3} \text{ s}\right) = 38.9 \text{ m/s}$$

2.25 (a)

$a = ?$ m/s^2 $a = ?$ m/s^2

$v_0 = 0$ m/s $v_0 = 30.0$ cm/s

$t_0 = 0$ s $t = ?$ s

$x_0 = 0$ m $x = 1.80$ cm

(b) GIVEN:

"Accelerated from rest" $\Rightarrow$ $v_0 = 0$ m/s.

"to 30.0 cm/s" $\Rightarrow$ $v = 0.300$ m/s.

"in a distance of 1.80 cm" $\Rightarrow$ $x - x_0 = 0.0180$ m.

(c) "How long" tells us to find t. To determine which equation to use, we look for an equation that has v_0, v, $x - x_0$, and t, since those are the parameters that we know or want to know. So, using Equations 2.7 and 2.8 gives us: $x = x_0 + \bar{v}\, t$ and $\bar{v} = \dfrac{v_0 + v}{2}$, or $x - x_0 = \left(\dfrac{v_0 + v}{2}\right) t$

Solving for t gives:

$$t = \frac{2(x - x_0)}{v_0 + v} = \frac{2(0.0180 \text{ m})}{(0 \text{ m/s}) + (0.300 \text{ m/s})} = \boxed{0.120 \text{ s}}$$

So it takes 120 ms to accelerate the blood from rest to 30.0 cm/s. Converting everything to standard units first, makes it easy to see that the units of meters cancel leaving only the units of seconds.

(d) Yes, the answer is reasonable. An entire heartbeat cycle takes about one second. The time for acceleration of the blood out of the ventricle is only a fraction of the entire cycle.

2.31 (a) FIND:

a (which should be negative).

GIVEN: "comes to a stop" $\Rightarrow$ $v = 0$ m/s.

"from an initial velocity of" $\Rightarrow$ $v_0 = 0.600$ m/s.

"in a distance of 2.00 m" $\Rightarrow$ $x - x_0 = 2.00 \times 10^{-3}$ m.

So, we need an equation that involves: a, v, v_0, and $x - x_0$, or we need Equation 2.11:

$$v^2 = v_0^2 + 2a(x - x_0), \text{ so that } a = \frac{v^2 - v_0^2}{2(x - x_0)} = \frac{(0 \text{ m/s})^2 - (0.600 \text{ m/s})^2}{2(2.00 \times 10^{-3} \text{ m})} = -90.0 \text{ m/s}^2.$$

So that the $\boxed{\text{deceleration} = 90.0 \text{ m/s}^2}$. To get the deceleration in multiples of g, we divide a by g:

$$\frac{a}{g} = \frac{-90.0 \text{ m/s}^2}{-9.80 \text{ m/s}^2} \Rightarrow \boxed{a = 9.18\, g}$$

(b) "Calculate the stopping time" they want us to find t. Using Equations 2.7 and 2.8 gives:

$$x - x_0 = \frac{1}{2}(v_0 + v)\, t, \text{ so that } t = \frac{2(x - x_0)}{v_0 + v} = \frac{2(2.00 \times 10^{-3} \text{ m})}{(0.600 \text{ m/s}) + (0 \text{ m/s})} = \boxed{6.67 \times 10^{-3} \text{ s}}$$

(c) To calculate the deceleration of the brain, use $x - x_0 = 4.50 \text{ mm} = 4.50 \times 10^{-3}$ m instead of 2.00 mm. Again, we use Equation 2.11, so that: $a = \dfrac{v^2 - v_0^2}{2(x - x_0)} = \dfrac{(0 \text{ m/s})^2 - (0.600 \text{ m/s})^2}{2(4.50 \times 10^{-3} \text{ m})} = -40.0 \text{ m/s}^2.$

And in expressed in multiples of g gives:

$$\frac{a}{g} = \frac{-40.0 \text{ m/s}^2}{-9.80 \text{ m/s}^2} \Rightarrow \boxed{a = 4.08\, g}$$

2.37 For falling objects, $a = -g = -9.80 \text{ m/s}^2$. And, from the problem statement we know that: $y_0 = 0 \text{ m}$ and $v_0 = +15.0 \text{ m/s}$. We want to find the displacement (y) and the velocity (v) at different times. *Remember that we are taking UP to be positive, so a is negative because gravity pulls down.* To find the displacement, we use Equation 2.10: $y = y_0 + v_0 t + \frac{1}{2}at^2$ and to find the velocity, we use Equation 2.9: $v = v_0 + at$.

For $t_1 = 0.500 \text{ s}$: $y_1 = y_0 + v_0 t_1 + \frac{1}{2}at_1^2 = 0 \text{ m/s} + (15.0 \text{ m/s})(0.500 \text{ s}) + \frac{1}{2}(-9.80 \text{ m/s}^2)(0.500 \text{ s})^2 = \boxed{6.28 \text{ m}}$

$v_1 = v_0 + at_1 = (15.0 \text{ m/s}) + (-9.80 \text{ m/s}^2)(0.500 \text{ s}) = \boxed{10.1 \text{ m/s}}$, the snowball is rising because $v>0$

For $t_2 = 1.00 \text{ s}$: $y_2 = y_0 + v_0 t_2 + \frac{1}{2}at_2^2 = 0 \text{ m/s} + (15.0 \text{ m/s})(1.00 \text{ s}) + \frac{1}{2}(-9.80 \text{ m/s}^2)(1.00 \text{ s})^2 = \boxed{10.1 \text{ m}}$

$v_2 = v_0 + at_2 = (15.0 \text{ m/s}) + (-9.80 \text{ m/s}^2)(1.00 \text{ s}) = \boxed{5.20 \text{ m/s}}$, the snowball is rising, but slowing down

For $t_3 = 1.50 \text{ s}$: $y_3 = y_0 + v_0 t_3 + \frac{1}{2}at_3^2 = 0 \text{ m/s} + (15.0 \text{ m/s})(1.50 \text{ s}) + \frac{1}{2}(-9.80 \text{ m/s}^2)(1.50 \text{ s})^2 = \boxed{11.5 \text{ m}}$

$v_3 = v_0 + at_3 = (15.0 \text{ m/s}) + (-9.80 \text{ m/s}^2)(1.50 \text{ s}) = \boxed{0.300 \text{ m/s}}$, the snowball is almost at the top.

For $t_4 = 2.00 \text{ s}$: $y_4 = y_0 + v_0 t_4 + \frac{1}{2}at_4^2 = 0 \text{ m/s} + (15.0 \text{ m/s})(2.00 \text{ s}) + \frac{1}{2}(-9.80 \text{ m/s}^2)(2.00 \text{ s})^2 = \boxed{10.4 \text{ m}}$

$v_4 = v_0 + at_4 = (15.0 \text{ m/s}) + (-9.80 \text{ m/s}^2)(2.00 \text{ s}) = \boxed{-4.60 \text{ m/s}}$, the snowball has begun to drop, and its speed will increase as it returns to the ground.

2.43 (a)

FIND: y_0, the initial height.

GIVEN: "it takes" $\Rightarrow$ $t = 2.35 \text{ s}$

"hit the ground" $\Rightarrow$ $y = 0 \text{ m}$

"initial velocity of ..." $\Rightarrow$ $v_0 = +8.00 \text{ m/s}$

"falling objects" $\Rightarrow$ $a = -g = -9.80 \text{ m/s}^2$.

What formula uses y_0, t, y, v_0, and a? Equation 2.10:

$$y = y_0 + v_0 t + \frac{1}{2}at^2, \text{ or } y_0 = y - v_0 t - \frac{1}{2}at^2$$

$y_0 = (0 \text{ m}) - (+8.00 \text{ m/s})(2.35 \text{ s}) - (0.5)(-9.80 \text{ m/s}^2)(2.35 \text{ s})^2 = \boxed{8.26 \text{ m}}$, so the cliff is 8.26 m high.

(b) GIVEN: Using the result from part (a) $\Rightarrow$ $y_0 = 8.26 \text{ m}$

The ball is "thrown straight DOWN with the same speed" $\Rightarrow$ $v_0 = -8.00 \text{ m/s}$

"falling objects" $\Rightarrow$ $a = -g = -9.80 \text{ m/s}^2$

FIND: t. What formula uses y_0, y, v_0, a and t? Equation 2.10, again! $y = y_0 + v_0 t + \frac{1}{2}at^2$

But this time we need to rewrite it in order to solve for t, which requires the quadratic formula, Equation 1.23:

$$\frac{1}{2}at^2 + v_0 t + (y_0 - y) = 0, \text{ so that } t = \frac{-v_0 \pm \sqrt{v_0^2 - 4(0.5a)(y_0 - y)}}{2(0.5a)}, \text{ or}$$

$$t = \frac{-(-8.00 \text{ m/s}) \pm \sqrt{(-8.00 \text{ m/s})^2 - 4(0.5)(-9.80 \text{ m/s}^2)(8.26 \text{ m} - 0 \text{ m})}}{2(0.5)(-9.80 \text{ m/s}^2)} = \frac{8.00 \text{ m/s} \pm 15.03 \text{ m/s}}{-9.80 \text{ m/s}^2}$$

$t = 0.717 \text{ s or } -2.35 \text{ s} \Rightarrow t = \boxed{0.717 \text{ s}}$, we only keep the positive value for the time, since the ball must strike the ground AFTER it was released.

Note: For problems **2.49** and **2.55**, all values are estimates from the graphs.

2.49 (a) In Figure 2.15(a), if we draw a tangent to the curve at $t = 20$ s, we can identify two points: $x = 0$ m, $t = 5$ s, and $x = 1500$ m, $t = 20$ s, so we can calculate an approximate slope:

$$v = \frac{\text{rise}}{\text{run}} = \frac{(1500 - 0)\,\text{m}}{(20 - 5)\,\text{s}} = 100 \text{ m/s} \approx 115 \text{ m/s}.$$

So, the slope of the displacement vs. time curve is the velocity curve.

(b) In Figure 2.15(b), we can identify two points: $v = 65$ m/s, $t = 10$ s, and $v = 140$ m/s, $t = 25$ s. Therefore, the slope is

$$a = \frac{\text{rise}}{\text{run}} = \frac{(140 - 65)\,\text{m/s}}{(25 - 10)\,\text{s}} = 5 \text{ m/s}^2 .$$

The slope of the velocity vs. time curve is the acceleration curve.

2.55 (a) In Figure 2.22(b), the slope at $t = 2.5$ s is horizontal, so the slope is zero:

$$a = \frac{\text{rise}}{\text{run}} = \frac{0 \text{ m/s}}{2.5 \text{ s}} = \boxed{0 \text{ m/s}^2 \text{ at } t = 2.5 \text{ s}}$$

(b) In Figure 2.22(b), drawing a tangent at $t = 5.0$ s, we can identify two points: $v = 0$ m/s, $t = 5$ s, and $v = -4$ m/s at $t = 6$ s, so

$$a = \frac{\text{rise}}{\text{run}} = \frac{\left[(-4) - 0\right]\text{m/s}}{(6 - 5)\,\text{s}} = \boxed{-4 \text{ m/s}^2 \text{ at } t=5.0 \text{ s}} .$$

At $t = 15$s, we can identify two points: $v = 0$ m/s at $t = 14$ s, and $v = 2$ m/s at $t = 15$ s, so

$$a = \frac{\text{rise}}{\text{run}} = \frac{(2 - 0)\,\text{m/s}}{(15 - 14)\,\text{s}} = \boxed{2 \text{ m/s}^2 \text{ at } t=15 \text{ s}}$$

Checking Figure 2.22(c), we read the acceleration from the graph to be 0 m/s^2, -4 m/s^2, and 2 m/s^2 at times $t = 2.5$ s, 5.0 s, and 15 s, respectively. This is consistent with our calculations.

TWO-DIMENSIONAL KINEMATICS ③

CONCEPTUAL QUESTIONS

3.1 A nonzero force is the only vector in the list. A nonzero force has a magnitude and a direction.

3.4 While driving a car your frame of reference is either the car or the ground. While flying in a commercial jet airplane your frame of reference is either the airplane or the ground. Instinctively, we use the ground a our frame of reference, since we are used to it being stationary.

3.7 Since the pilot was not told which direction to fly, she could end up anywhere on the circle and still be 123 km from San Francisco. She would need to be told to fly 45° N of E, or due NNE, in order to get to Sacramento.

3.10 The greatest magnitude is produced when the vectors point in the same relative direction, producing a vector of length $A + B$. The minimum magnitude is produced when the vectors point in the opposite relative direction, producing a vector of length $|A - B|$.

3.13 If the vectors are perpendicular, then the component of **A** along the direction of **B** is zero, and the component of **B** along the direction of **A** is zero, since the angle between them is $90°$.

3.16 Air resistance will decrease the range and maximum height of a projectile.

3.19 The quarter that was flicked off the edge of the table will follow a parabolic path downward, while the quarter that was nudged will fall basically straight downward to the floor. Both quarters will land on the floor at precisely the same time (neglecting air resistance).

3.22 Relative to the truck, the clod of dirt that hits the ground directly below the end of the truck has a velocity in the vertical (downward) direction only. Relative to the ground, however, it will have a velocity downward as seen by the truck plus a forward velocity equal to the velocity of the truck. When the clod leaves the truck, it has only a forward velocity of the speed of the truck and gravity is what pulls it downward. According to the truck, the clod falls straight down; according to the ground, the clod follows a parabolic path downward.

PROBLEMS

3.1 (a) To measure the total distance traveled, we take a ruler and measure the length of Path A to the north, and add to it the length of Path A to the east. Path A travels 3 blocks north and 1 block east, for a total of 4 blocks.
Each block is 120 m, so the distance traveled is: $d = (4 \times 120 \text{ m}) = \boxed{480 \text{ m}}$

(b) Graphically, measure the length and angle of the line from the start to the arrow of Path A. Use a protractor to measure the angle, with the center of the protractor at the start, measure the angle to where the arrow is at the end of Path A. In order to do this, it may be necessary to extend the line from the start to the arrow of Path A, using a ruler. The length of the displacement vector, measured from the start to the arrow of Path A, along the line you just drew. $\vec{s} = \boxed{379 \text{ m, } 18.4° \text{ east of north}}$

3.7   (a)

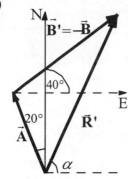

To do this problem, draw the two vectors $\vec{A}$ and $\vec{B}' = -\vec{B}$ tip to tail as shown at the right.

The vector $\vec{A}$ should be 12.0 units long and at an angle of $20°$ to the left of the y-axis. Then at the arrow of vector $\vec{A}$, draw the vector $\vec{B}' = -\vec{B}$, which should be 20.0 units long and at an angle of $40°$ above the x-axis.

The resultant vector, $\vec{R}'$, goes from the tail of vector A to the tip of vector B, and therefore has an angle of α above the x-axis.

Measure the length of the resultant vector using your ruler, and use a protractor with the center at the tail of the resultant vector to get the angle.

$$R' = \boxed{26.6 \text{ m}} \text{, and } \alpha = \boxed{65.1° \text{ north of east}}$$

(b)

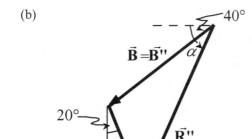

To do this problem, draw the two vectors $\vec{B}$ and $\vec{A}'' = -\vec{A}$ tip to tail as shown at the right.

The vector $\vec{B}$ should be 20.0 units long and at an angle of $40°$ below the $-x$-axis.

Then at the arrow of vector $\vec{B}$, draw the vector $\vec{A}'' = -\vec{A}$, which should be 12.0 units long and at an angle of $20°$ to the right of the negative y-axis.

The resultant vector, $\vec{R}'$, goes from the tail of vector $\vec{B}$ to the tip of vector $\vec{A}''$, and therefore has an angle of α below the $-x$-axis.

Measure the length of the resultant vector using your ruler, and use a protractor with the center at the tail of the resultant vector to get the angle.

$$R'' = \boxed{26.6 \text{ m}} \text{, and } \alpha = \boxed{65.1° \text{ south of west}}$$

So the length is the same, but the direction is reversed from part (a).

3.13   (a)   To solve this problem analytically, add up the distance by counting the blocks traveled along the line that is Path C:

$$d = (1 \times 120 \text{ m}) + (5 \times 120 \text{ m}) + (2 \times 120 \text{ m}) + (1 \times 120 \text{ m}) + (1 \times 120 \text{ m}) + (3 \times 120 \text{ m}) = \boxed{1.56 \times 10^3 \text{ m}}$$

(b)   To get the displacement, calculate the displacement in the x- and y-directions separately, then use the formulas for adding vectors covered on page 62: The displacement in the x-direction is calculated by adding the x-distance traveled in each leg, being careful to subtract values when they are negative:

$$s_x = (0 + 600 + 0 - 120 + 0 - 360) \text{ m} = 120 \text{ m} .$$

Using the same method, calculate the displacement in the y-direction:

$$s_y = (120 + 0 - 240 + 0 + 120 + 0) \text{ m} = 0 \text{ m} .$$

Now, using Equations 3.7 and 3.8, calculate the total displacement vectors:

$$s = \sqrt{s_x^2 + s_y^2} = \sqrt{(120 \text{ m})^2 + (0 \text{ m})^2} = 120 \text{ m};$$

$$\theta = \tan^{-1}(s_y/s_x) = \tan^{-1}(0 \text{ m}/120 \text{ m}) = 0° \Rightarrow \text{ east}$$

so that

$$\vec{s} = \boxed{120 \text{ m, East}}$$

3.19 (a)

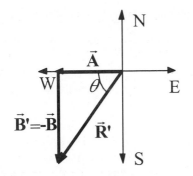

We want to calculate the displacement for a walk 18.0 m to the west, followed by 25.0 m to the south. First, calculate the displacement in the x- and y-directions, using Equations 3.5 and 3.6: (The angles are measured from due east)

$$R_x = -18.0 \text{ m}, \quad R_y = -25.0 \text{ m}$$

Then, using Equations 3.7 and 3.8, calculate the total displacement vectors: $R' = \sqrt{R_x^2 + R_y^2} = \sqrt{(18.0 \text{ m})^2 + (25.0 \text{ m})^2} = \boxed{30.8 \text{ m}}$

$$\theta = \tan^{-1} \frac{\text{opp}}{\text{adj}} = \tan^{-1}\left(\frac{25.0 \text{ m}}{18.0 \text{ m}}\right) = \boxed{54.2° \text{ south of west}}$$

(b)

Now do the same calculation, except walk 25.0 m to the north, followed by 18.0 m to the east. Equations 3.5 and 3.6 give:

$$R_x = 18.0 \text{ m}, \quad R_y = 25.0 \text{ m}$$

Then, using Equations 3.7 and 3.8 gives:

$$R'' = \sqrt{R_x^2 + R_y^2} = \sqrt{(18.0 \text{ m})^2 + (25.0 \text{ m})^2} = \boxed{30.8 \text{ m}}$$

$$\phi = \tan^{-1} \frac{\text{opp}}{\text{adj}} = \tan^{-1}\left(\frac{25.0 \text{ m}}{18.0 \text{ m}}\right) = \boxed{54.2 \text{ N of E}}.$$

Which is consistent with part (a)!

3.25

Start with the information given: $v_{\text{tot}} = 6.72$ m/s; 49° N of E .
We also know: v_A is 22.5° N of E . To calculate the angle of v_B, first notice that $\alpha + \beta = 180°$, since $\vec{v}_A$ and the dashed line form a straight line.
Then, since the internal angles of a triangle add up to 180° , we know that

$$\beta + 26.5° + 23.0° = 180°.$$

Now, solving for β in the second equation gives:

$$\beta = 180° - 26.5° - 23.0° = 180° - 49.5°.$$

Finally, substituting into the first equation, we can solve for α :

$$\alpha = 180° - \beta = 180° - (180° - 49.5°) = 49.5°$$

so that v_B is at an angle of $\alpha + \phi = 72.0°$ N of E .

Now, getting the components of all vectors gives:

$v_{\text{tot},x} = (6.72 \text{ m/s})\cos 49.0° = 4.409 \text{ m/s}, \quad v_{\text{tot},y} = (6.72 \text{ m/s})\sin 49.0° = 5.072 \text{ m/s};$

$v_{A,x} = v_A \cos 22.5°, \quad v_{A,y} = v_A \sin 22.5°; \quad v_{B,x} = v_B \cos 72.0°, \quad v_{B,y} = v_B \sin 72.0°$

(1): $v_{\text{tot }x} = 4.409 \text{ m/s} = v_A \cos 22.5° + v_B \cos 72.0° = v_A(0.9239) + v_B(0.3090)$

(2): $v_{\text{tot }y} = 5.072 \text{ m/s} = v_A \sin 22.5° + v_B \sin 72.0° = v_A(0.3827) + v_B(0.9511)$

Dividing equation (1) by (0.9239) gives: $v_A = 4.772 \text{ m/s} - (0.3345)v_B$. Substituting into equation (2) gives:

$$\left[4.772 \text{ m/s} - (0.3345)v_B\right](0.3827) + v_B(0.9511) = 5.072 \text{ m/s}, \text{ or}$$

$$\boxed{v_B = 3.94 \text{ m/s and } v_A = 3.45 \text{ m/s}}$$

3.31 (a) Equation 3.13 gives the range of a projectile on *level ground* for which air resistance is negligible:

$R = \dfrac{v_0^2 \sin 2\theta_0}{g}$, where v_0 is the initial speed and θ_0 is the initial angle relative to the horizontal. Solving for

the initial angle gives: $\theta_0 = \dfrac{1}{2}\sin^{-1}\left(\dfrac{gR}{v_0^2}\right)$, where: $R = 25.0$ m, $v_0 = 20.0$ m/s, and $g = 9.80$ m/s^2. Therefore:

$$\theta_0 = \frac{1}{2}\sin^{-1}\left[\frac{(9.80 \text{ m/s})(25.0 \text{ m})}{(20.0 \text{ m/s})^2}\right] = \boxed{18.9°}$$

(b) Looking at Equation 3.13, we see that the range will be the same for another angle, θ_0', where $\theta_0 + \theta_0' = 90°$

or $\theta_0' = 90° - \theta_0 = 90° - 18.89° = \boxed{71.1°}$. This angle is not used as often, because the time of flight will be longer. In football that means the defense would have a greater time to get into position to knock down or intercept the pass that has the larger angle of release.

3.37

x-Direction (Horizontal)	*y*-Direction (Vertical)

Given: $v_{0,x} = 3.00$ m/s   Given: $v_{0,y} = 0.00$ m/s

$a_x = 0$ m/s^2 $a_y = -g = -9.80$ m/s^2

$\Delta y = (y - y_0) = -5.00$ m

From Equation 3.9b: Use Equation 3.10d to calculate v_y : $v_y^2 = v_{0,y}^2 - 2g(y - y_0)$

$v_x = v_{0,x} = $ constant $= 3.00$ m/s

$$v_y = \sqrt{(0 \text{ m/s})^2 - 2(9.80 \text{ m/s}^2)(-5.00 \text{ m})} = -9.90 \text{ m/s}$$

Now we can calculate the final velocity, using Equations 3.11c and 3.11d:

$$v = \sqrt{v_x^2 + v_y^2} = \sqrt{(3.00 \text{ m/s})^2 + (-9.90 \text{ m/s})^2} = 10.3 \text{ m/s, and}$$

$$\theta = \tan^{-1}\left(\frac{v_y}{v_x}\right) = \tan^{-1}\left(\frac{-9.90 \text{ m/s}}{3.00 \text{ m/s}}\right) = -73.1°$$

so that:

$$\boxed{\vec{v} = 10.3 \text{ m/s, } 73.1° \text{ below the horizontal}}$$

3.43 (a) Given: $v_x = 5.00$ m/s, $y - y_0 = 0.75$ m, $v_y = 0$ m/s, $a_y = -g = -9.80$ m/s^2. Find: $v_{0,y}$.

Using Equation 3.10d, $v_y^2 = v_{0,y}^2 - 2g(y - y_0)$, gives:

$$v_{0,y} = \sqrt{v_y^2 + 2g(y - y_0)} = \sqrt{(0 \text{ m/s})^2 + 2(9.80 \text{ m/s}^2)(0.75 \text{ m})} = \boxed{3.83 \text{ m/s}}$$

(b) To calculate the *x*-direction information, remember that the time is the same in the *x*- and *y*-directions. Calculate the time from the *y*-direction information, then use it to calculate the *x*-direction information: Use Equation 3.10b to calculate the time:

$$v_y = v_{0,y} - gt, \text{ so that } t = \frac{v_{0,y} - v_y}{g} = \frac{(3.83 \text{ m/s}) - (0 \text{ m/s})}{9.80 \text{ m/s}^2} = 0.391 \text{ s}.$$

Now, using Equation 3.9a gives the horizontal distance he travels to the basket:

$$x = x_0 + v_x t, \text{ so that } (x - x_0) = v_x t = (5.00 \text{ m/s})(0.391 \text{ s}) = \boxed{1.96 \text{ m}}.$$

So he must leave the ground 1.96 m before the basket to be at his maximum height when he reaches the basket.

3.49 (a) To keep track of the runners, let's label F for the first runner and S for the second runner. Then, we are given: $\vec{v}_F = 3.50$ m/s, and $\vec{v}_S = 4.20$ m/s . To calculate the velocity of the second runner relative to the first, subtract the velocity vectors:

$$\vec{v}_{SF} = \vec{v}_S - \vec{v}_F = 4.20 \text{ m/s} - 3.50 \text{ m/s} = \boxed{0.70 \text{ m/s}} \text{ (faster than first runner)}.$$

(b) Use the definition of velocity (Equation 2.3) to calculate the time for each runner separately. For the first runner, she runs 250 m with at a velocity of 3.50 m/s:

$$t_F = \frac{x_F}{v_F} = \frac{250 \text{ m}}{3.50 \text{ m/s}} = 71.43 \text{ s}$$

For the second runner, she runs 45 m farther than the first runner at a velocity of 4.20 m/s:

$$t_S = \frac{x_S}{v_S} = \frac{250 \text{ m} + 45 \text{ m}}{4.20 \text{ m/s}} = 70.24 \text{ s}.$$

So, since $t_S < t_F$, the $\boxed{\text{second runner will win}}$.

(c) We can calculate their relative position, using their relative velocity and the time of travel. Initially, the second runner is 45 m behind, the relative velocity was found in part (a), and the time is the time for the second runner, so:

$$x_{SF} = x_{0,SF} + v_{SF} t_S = -45.0 \text{ m} + (0.70 \text{ m/s})(70.24 \text{ s}) = \boxed{4.17 \text{ m}}$$

3.55

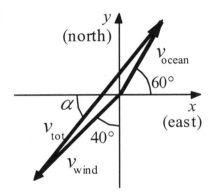

In order to calculate the velocity of the wind relative to the ocean, we need to add the vectors for the wind and the ocean together, being careful to use vector addition. The velocity of the wind relative to the ocean is equal to the velocity of the wind relative to the earth plus the velocity of the earth relative to the ocean. Now,

$$\vec{v}_{W,O} = \vec{v}_{W,E} + \vec{v}_{E,O} = \vec{v}_{W,E} - \vec{v}_{O,E}$$

The first subscript is the object, the second is what it is relative to, so that $\vec{v}_{AB} = -\vec{v}_{BA}$, in other words, the velocity of the earth relative to the ocean is the opposite of the velocity of the ocean relative to the earth.

To solve this vector equation, we need to add the x- and y-components separately.

$$v_{WO,x} = v_{WE,x} - v_{OE,x} = (-4.50 \text{ m/s})\cos 50° - (2.20 \text{ m/s})\cos 60° = -3.993 \text{ m/s}$$

$$v_{WO,y} = v_{WE,y} - v_{OE,y} = (-4.50 \text{ m/s})\sin 50° - (2.20 \text{ m/s})\sin 60° = -5.352 \text{ m/s}$$

Finally, we can use Equations 3.16 and 3.17 to calculate the velocity of the water relative to the ocean:

$$v = \sqrt{v_x^2 + v_y^2} = \sqrt{(-3.993 \text{ m/s})^2 + (-5.352 \text{ m/s})^2} = \boxed{6.68 \text{ m/s}},$$

$$\alpha = \tan^{-1} \frac{|v_y|}{|v_x|} = \tan^{-1}\left(\frac{5.352 \text{ m/s}}{3.993 \text{ m/s}}\right) = \boxed{53.3° \text{ south of west}}.$$

3.61

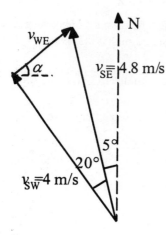

To calculate the velocity of the water relative to the earth, we need to add the vectors. The velocity of the water relative to the earth is equal to the velocity of the water relative to the ship plus the velocity of the ship relative to the earth.

$$\vec{v}_{WE} = \vec{v}_{WS} + \vec{v}_{SE} = -\vec{v}_{SW} + \vec{v}_{SE}$$

Now, we need to calculate the x- and y-components separately:

$$v_{WE,x} = -v_{SW,x} + v_{SE,x} = -(4.00 \text{ m/s})\cos(115)° + (4.80 \text{ m/s})\cos(95)° = 1.272 \text{ m/s}$$

$$v_{WE,y} = -v_{SW,y} + v_{SE,y} = -(4.00 \text{ m/s})\sin(115)° + (4.80 \text{ m/s})\sin(95)° = 1.157 \text{ m/s}$$

Finally, we use Equations 3.16 and 3.17 to calculate the velocity of the water relative to the earth:

$$v_{WE} = \sqrt{v_{WE,x}^2 + v_{WE,y}^2} = \sqrt{(1.272 \text{ m/s})^2 + (1.157 \text{ m/s})^2} = \boxed{1.72 \text{ m/s}},$$

$$\alpha = \tan^{-1}\left(\frac{v_{WE,y}}{v_{WE,x}}\right) = \tan^{-1}\frac{1.157 \text{ m/s}}{1.272 \text{ m/s}} = \boxed{42.3° \text{ north of east}}.$$

DYNAMICS

4

CONCEPTUAL QUESTIONS

4.1 (b) is correct. Net forces cause changes in motion, not strictly motion. An object in motion will remain in motion unless acted upon by a net force. The net force changes the motion.

4.4 Another force standard could be to apply the force to a ball. If we apply the force directly upward to a ball of known mass we can measure the maximum height of the ball and therefore calculate the force applied.

4.7 Forces holding a body together are internal forces, and therefore all are balanced by equal and opposite forces. So, we can neglect internal forces when applying Newton's second law of motion.

4.10 An object floating in space, for instance, can have a nonzero velocity while no net external force is applied. As long as there are no external forces applied to the object, it will continue to float in a straight line at the same speed.

4.13 If the acceleration of a system is zero, that there can be no NET forces acting. That does not mean that there cannot be external or internal forces. As long as the external forces cancel, there can be external forces present. Also, internal forces always are present, since molecules in an object feel forces from neighboring molecules, etc., and will be present even if a system has a zero acceleration.

4.16

If the lineman pushes hard enough to generate a force that is larger than the frictional force the opposing player generates with the ground, then the lineman will make the opposing player move backwards. The equal and opposite forces are generated at the players hands, and are internal forces. It seems as though the lineman can never outpush his opponent, when considering just his hands, but when you consider the system of the two players together, the hands are internal forces, while the forces at the ground are external and can be unbalanced.

4.19 A normal force is a force present when two objects are in contact. Then normal force is perpendicular to the contact surface between the two objects. Simple friction is proportional to the normal force, and points in a direction that opposes the motion of the object.

4.22 The chalk screeches when you are reversing direction or making a tight curve so that it momentarily stops. When it stops you must overcome static friction to get it moving again. Since static friction is greater than kinetic friction, it takes more force to start it moving again, and so the chalk jerks and screeches.

4.25 Child #1 has a toy, which child #2 wants. Child #2 takes the toy from child #1 (exchanging the particle). Child #1 comes running at child #2, trying to get the toy back. This attractive force, bringing child #1 back toward child #2, occurs after the toy has been transferred.

PROBLEMS

4.1   The net force acting on the sprinter is given by Equation 4.2:

$$\text{net } F = ma = (63.0 \text{ kg})(4.20 \text{ m/s}^2) = \boxed{265 \text{ N}}$$

4.7   (a) Following Example 4.2, with only one rocket burning, net $F = T - f$ so that Newton's second law gives:

$$a = \frac{\text{net } F}{m} = \frac{T - f}{m} = \frac{2.59 \times 10^4 \text{ N} - 600 \text{ N}}{2100 \text{ kg}} = \boxed{12.0 \text{ m/s}^2}$$

(b) To reduce the acceleration by a factor of four, the friction would also have to be reduced by a factor of four.

4.13   *Step 1.*
     Use
     Newton's
     Laws of
     Motion.

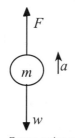

Forces acting on
the high jumper.

Step 2. Given: $a = 4.00g = (4.00)(9.80 \text{ m/s}^2) = 39.2 \text{ m/s}^2$; $m = 70.0$ kg
     Find: F .

Step 3. net $F = +F - w = ma$, so that $F = ma + w = ma + mg = m(a + g)$

$$F = (70.0 \text{ kg})\left[(39.2 \text{ m/s}^2) + (9.80 \text{ m/s}^2)\right] = \boxed{3.43 \times 10^3 \text{ N}}$$

The force exerted by the high-jumper is actually down on the ground, but F is up from the ground to help him jump.

Step 4. This result is reasonable, since it is quite possible for a person to exert a force of the magnitude of 10^3 N .

4.19   (a) Since F_2 is the *y*-component of the total force:

$$F_2 = F_{\text{tot}} \sin 35° = (20 \text{ N})\sin 35° = 11.47 \text{ N} = \boxed{11 \text{ N}} .$$

And F_1 is the x-component of the total force:

$$F_{\text{tot}} \cos 35° = F_1 = (20 \text{ N})\cos 35° = 16.38 \text{ N} = \boxed{16 \text{ N}} .$$

(b)

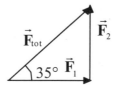

is the same as:

(c)

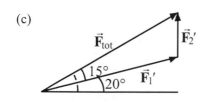

$F_1' \cos 20° = F_1$

$$F_1' = \frac{F_1}{\cos 20°} = \frac{16.38 \text{ N}}{\cos 20°} = 17.4 \text{ N} = \boxed{17 \text{ N}}$$

$$F_1' \sin 20° + F_2' = F_2 \Rightarrow F_2' = F_2 - F_1' \sin 20° = \boxed{5.53 \text{ N}}$$

(Note: numerous other possibilities are also possible)

4.25   The weight on the moon is equal to the mass time the gravity on the moon. The weight on earth is equal to the mass times the gravity on the earth. The **mass** is unchanged, while the **weight** depends on gravity.

$w_m = mg_m$; $w_e = mg_e$, so setting the mass equal gives:

$$m = \frac{w_m}{g_m} = \frac{w_e}{g_e} \Rightarrow w_e = w_m \frac{g_e}{g_m} = \left(60.0 \text{ lb}\right)\frac{9.80 \text{ m/s}^2}{1.67 \text{ m/s}^2} = \boxed{352 \text{ lb}}$$

4.31   (a)   Given: $F_D = 7.50 \times 10^4$ N ; $m = 5.00 \times 10^6$ kg. Find: μ_k. We know that the frictional force is equal to the coefficient of friction time the normal force. The normal force, in this case, is equal to the weight force.

$$N = w = mg = \left(5.00 \times 10^6 \text{ kg}\right)\left(9.80 \text{ m/s}^2\right) = 4.90 \times 10^7 \text{ N} ,$$

so that:

$$F_D = \mu_k N \Rightarrow \mu_k = \frac{F_D}{N} = \frac{7.50 \times 10^4 \text{ N}}{4.90 \times 10^7 \text{ N}} = \boxed{1.53 \times 10^{-3} \text{ (unitless)}}$$

(b)   This is consistent with the small amount of drag experience by ships. Barges are designed to experience very small amounts of drag when traveling at low speeds.

4.37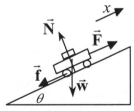

The component of $\vec{w}$ down the incline leads to the acceleration:

$$\text{net } F_x = ma = mg \sin\theta \text{ so that } \boxed{a = g \sin\theta}$$

The component of $\vec{w}$ perpendicular to the incline equals the normal force. net $F_y = 0 = N - mg \cos\theta$

4.43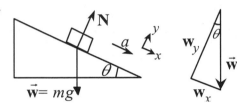

Take the x-direction as positive up the slope.
net $F_x = F - f - w_x$, and $w = mg$

"Only half the weight of the car..." $\Rightarrow F = \frac{1}{2}\mu_s w_y = \frac{1}{2}\mu_s w \cos\theta$

"No slipping" $\Rightarrow f = \mu_k w_y = 0$

Substituting gives:

$$\text{net } F_x = F - w_x = \frac{1}{2}\mu_s mg \cos\theta - mg \sin\theta = ma$$

So the maximum acceleration is:

$$a = g\left(\frac{1}{2}\mu_s \cos\theta - \sin\theta\right)$$

Now, applying the specific values of μ_s gives:

(a)   On dry concrete, $\mu_s = 1.00$, so that: $a = \left(9.80 \text{ m/s}^2\right)\left[\frac{1}{2}(1.00)\cos 4° - \sin 4°\right] = \boxed{4.20 \text{ m/s}^2}$

(b)   On wet concrete, $\mu_s = 0.700$, so that: $a = \left(9.80 \text{ m/s}^2\right)\left[\frac{1}{2}(0.700)\cos 4° - \sin 4°\right] = \boxed{2.74 \text{ m/s}^2}$

(c)   On ice, $\mu_s = 0.100$, so that: $a = \left(9.80 \text{ m/s}^2\right)\left[\frac{1}{2}(0.100)\cos 4° - \sin 4°\right] = \boxed{-0.195 \text{ m/s}^2}$ Since the acceleration is negative, that tells us that the car cannot make it up the grade.

4.49 *Step 1*: Use Newton's Laws since
we are looking for forces.

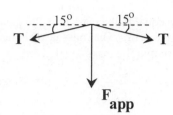

Step 2: Draw a free body diagram:

Step 3: Given: $T = 25.0\text{ N}$. Find F_{app}. Using Newton's Laws gives: net $F_y = 0$, so that the applied force is due to the *y*-components of the two tensions:
$$F_{app} = 2T\sin\theta = 2(25.0\text{ N})\sin 15° = 12.9\text{ N}.$$
The *x*-components of the tensions cancel.
$$\text{net } F_x = 0$$

Step 4: This seems reasonable, since the applied tensions should be greater than the force applied to the tooth.

4.55 (a)

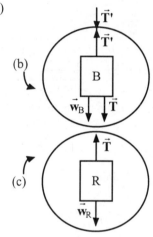

(b) Using the upper circle of the diagram,
$$\text{net } F_y = 0, \text{ so that } T' - T - w_B = 0.$$
Using the lower circle of the diagram,
$$\text{net } F_y = 0, \text{ giving } T - w_R = 0.$$
Next, write the weights in terms of masses:
$$w_B = m_B g,\ w_R = m_R g.$$
Solving for the tension in the upper rope gives:
$$T' = T + w_B = w_R + w_B = m_R g + m_B g = g(m_R + m_B)$$
Plugging in the numbers gives:
$$T' = (9.80\text{ m/s}^2)(55.0\text{ kg} + 90.0\text{ kg}) = \boxed{1.42 \times 10^3\text{ N}}$$

(c) Using the lower circle of the diagram, net $F_y = 0$, so that $T - w_R = 0$. Again, write the weight in terms of mass: $w_R = m_R g$ Solving for the tension in the lower rope gives:
$$T = m_R g = (55.0\text{ kg})(9.80\text{ m/s}^2) = \boxed{539\text{ N}}$$

4.61 (a, d)

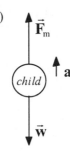

(a) The net force exerted on the child must equal the mass times the acceleration:
$$\text{net } F = ma = F_m - w$$
The weight is given by $w = mg$, so the force exerted by the mother is:
$$F_m = ma + w = ma + mg = m(a + g).$$
The acceleration is $a = 0.850 \text{ m/s}^2$, and plugging in the numbers gives:
$$F_m = (12.0 \text{ kg})(0.850 \text{ m/s}^2 + 9.80 \text{ m/s}^2) = \boxed{128 \text{ N}}$$

(d) To get the ratio, simply divide by the weight:
$$\frac{F_m}{w} = \frac{m(a+g)}{mg} = \frac{a+g}{g} = \frac{0.850 \text{ m/s}^2 + 9.80 \text{ m/s}^2}{9.80 \text{ m/s}^2} = \boxed{1.09}$$

(b, d)

(b) The net force exerted on the child equals the mass times the acceleration, or zero:
$$\text{net } F = ma = F_m - w = 0.$$
So, the force exerted by the mother must equal the child's weight:
$$F_m = mg = (12.0 \text{ kg})(9.80 \text{ m/s}^2) = \boxed{118 \text{ N}}$$

(d) To get the ratio, divide by the weight: $\dfrac{F_m}{w} = \dfrac{mg}{mg} = \boxed{1.00}$

(c, d)

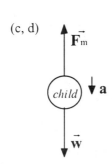

(c) The net force exerted equals the mass times the acceleration (which is downward):
$$\text{net } F = -ma = F_m - w$$
The force exerted by the mother is: $F_m = w - ma = mg - ma = m(g - a)$.
The acceleration is now $a = 2.30 \text{ m/s}^2$, giving:
$$F_m = (12.0 \text{ kg})(9.80 \text{ m/s}^2 - 2.30 \text{ m/s}^2) = \boxed{90.0 \text{ N}}$$

(d) To get the ratio, divide by the weight:
$$\frac{F_m}{w} = \frac{g-a}{g} = \frac{9.80 \text{ m/s}^2 - 2.30 \text{ m/s}^2}{9.80 \text{ m/s}^2} = \boxed{0.765}$$

4.67 (a) After he leaves the ground, the basketball player is like a projectile, so use Equation 3.10d. Since he reaches a maximum height of 0.900 m, $v^2 = v_0^2 - 2g(y - y_0)$, with $y - y_0 = 0.900 \text{ m}$, and $v = 0 \text{ m/s}$. Solving for the initial velocity gives:
$$v_0 = \sqrt{2g(y - y_0)} = \sqrt{2(9.80 \text{ m/s}^2)(0.900 \text{ m})} = \boxed{4.20 \text{ m/s}}$$

(b) Since we want to calculate his acceleration, use Equation 2.11: $v^2 = v_0^2 + 2a(y - y_0)$, where $y - y_0 = 0.300 \text{ m}$, and since he starts from rest, $v_0 = 0 \text{ m/s}$. Solving for the acceleration gives:
$$a = \frac{v^2}{2(y - y_0)} = \frac{(4.20 \text{ m/s})^2}{(2)(0.300 \text{ m})} = \boxed{29.4 \text{ m/s}^2}$$

(c)

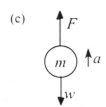

Now, we must draw a free body diagram, in order to calculate the force exerted by the basketball player to jump. The net force is equal to the mass times the acceleration:
$$\text{net } F = ma = F - w = F - mg$$
So, solving for the force gives: $F = ma + mg = m(a + g)$
$$F = 110 \text{ kg}(29.4 \text{ m/s}^2 + 9.80 \text{ m/s}^2) = \boxed{4.31 \times 10^3 \text{ N}}$$

4.73   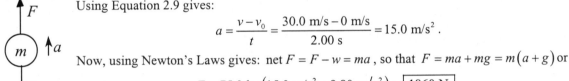The net force is due to the tension and the weight: net $F = ma = T - w = T - mg$, and $m = 1700$ kg.

(a) $a = 1.20$ m/s^2, so the tension is: $T = m(a + g) = (1700 \text{ kg})(1.20 \text{ m/s}^2 + 9.80 \text{ m/s}^2) = \boxed{1.87 \times 10^4 \text{ N}}$

(b) $a = 0$ m/s^2, so the tension is: $T = w = mg = (1700 \text{ kg})(9.80 \text{ m/s}^2) = \boxed{1.67 \times 10^4 \text{ N}}$

(c) $a = 0.600$ m/s^2, but *down*: $T = m(g - a) = (1700 \text{ kg})(9.80 \text{ m/s}^2 - 0.600 \text{ m/s}^2) = \boxed{1.56 \times 10^4 \text{ N}}$

(d)   Use Equations 2.10 and 2.9 (in the y-direction): $y - y_0 = v_0 t + \frac{1}{2}at^2$ and $v = v_0 + at$.

For part (a), $v_0 = 0$ m/s, $a = 1.20$ m/s^2, $t = 1.50$ s, giving

$y_1 = \frac{1}{2}a_1t_1^2 = \frac{1}{2}(1.20 \text{ m/s}^2)(1.50 \text{ s})^2 = 1.35$ m and $v_1 = a_1t_1 = (1.20 \text{ m/s}^2)(1.50 \text{ s}) = 1.80$ m/s.

For part (b), $v_0 = v = 1.80$ m/s, $a = 0$ m/s^2, $t = 8.50$ s, so

$$y_2 = v_1t_2 = (1.80 \text{ m/s})(8.50 \text{ s}) = 15.3 \text{ m}.$$

For part (c): $v_0 = 1.80$ m/s, $a = -0.600$ m/s^2, $t = 3.00$ s, so that:

$$y_3 = v_2t + \frac{1}{2}a_3t_3^2 = (1.80 \text{ m/s})(3.00 \text{ s}) + 0.5(-0.600 \text{ m/s}^2)(3.00 \text{ s})^2 = 2.70 \text{ m}$$

$$v_3 = v_2 + a_3t_3 = 1.80 \text{ m/s} + (-0.600 \text{ m/s}^2)(3.00 \text{ s}) = 0 \text{ m/s}$$

Finally, the total distance traveled is $y_1 + y_2 + y_3 = 1.35 \text{ m} + 15.3 \text{ m} + 2.70 \text{ m} = 19.35 \text{ m} = \boxed{19.4 \text{ m}}$

And the final velocity will be the velocity at the end of part (c), or $v = \boxed{0 \text{ m/s}}$

4.79   (a)   Using Equation 2.9 gives:

$$a = \frac{v - v_0}{t} = \frac{30.0 \text{ m/s} - 0 \text{ m/s}}{2.00 \text{ s}} = 15.0 \text{ m/s}^2.$$

Now, using Newton's Laws gives: net $F = F - w = ma$, so that $F = ma + mg = m(a + g)$ or

$$F = 75.0 \text{ kg}(15.0 \text{ m/s}^2 + 9.80 \text{ m/s}^2) = \boxed{1860 \text{ N}}.$$

The ratio of the force to the weight is then:

$$\frac{F}{w} = \frac{m(a + g)}{mg} = \frac{15.0 \text{ m/s}^2 + 9.80 \text{ m/s}^2}{9.80 \text{ m/s}^2} = \boxed{2.53}$$

(b)   The value (1860 N) is more force than you expect to experience on an elevator.

(c)   The acceleration $a = 15.0$ m/s$^2 = 1.53g$ is much higher than any standard elevator. The final speed is too large (30.0 m/s is VERY fast)! The time of 2.00 s is not unreasonable for an elevator.

STATICS, TORQUE, AND ELASTICITY

5

CONCEPTUAL QUESTIONS

5.1 When a system is in dynamic equilibrium, its velocity is nonzero, but constant. See Figure 5.1b for a diagram of a car in dynamic equilibrium, with all external forces labeled on it. Notice that all forces cancel, so that the net force on the car is zero and the car moves at a constant velocity.

5.4 If the wrecking ball hits near the top of the wall, the force exerted on the wall will cause at least some of the wall to move horizontally, in the direction of the force. Also, the force, being applied away from the fixed point, will create a torque, causing the wall to tend to rotate about its fixed point, the base. So when you hit it at the top the torque is large due to the long lever arm and the wall rotates about the base. The same effects will be present if the wrecking ball hits the wall near the bottom of the wall, but the torque generated will be much less, since the lever arm is smaller. So, if the force is applied near the base, it will tend to make the wall fall straight down, after the wrecking ball knocks out a hole.

5.7 The center of mass can indeed be located outside an object. For instance, consider a tire. The center of mass is located directly in the center of the tire, where there is no rubber present. The same is true of a horseshoe shaped object; the center of mass is outside the physical object. A system acts as if its mass is concentrated at the center of mass, but the center of mass does not need to be physically within the object.

5.10 When the rope is absolutely tight, the tightrope walker is in unstable equilibrium above the tightrope. A slightly loose tightrope allows the tightrope walker to make slight adjustments to the location of her center of mass relative to the position of the tightrope by moving her feet. This extra way of adjusting the location of her center of mass, makes tightrope walking quite a bit easier. With the ability to make adjustments to the position of the rope, the torque she generates with her feet is opposite to the direction of the displacement, and there is a range where she is in stable equilibrium, making her survival much higher.

5.13 In order to balance the load, its center of mass must be located over her center of mass. Since her center of mass is located along her neck vertebrae, it is necessary for the load to be located over her neck vertebrae also.

5.16 The nutcracker is most similar to a wheelbarrow, in Figure 5.16(a), because both have the input and output forces on the same side of the pivot and they have a larger output force than the input force applied.

5.19 Friction plays an important role in actual pulley systems and causes them to be less than ideal. The friction is present in the rotation of the pulley, and that is why pulleys need to be oiled to reduce this friction.

5.22 As we see in Figure 5.19, the force at the joint is almost at the point of rotation, whereas the external force exerted on the book is much farther from the point of rotation. Since the torque on the arm will be the same everywhere, the force produced at our joints is much larger than the external force because it has a smaller lever arm. These forces on the joints will be even larger than the muscle forces, discussed in Question 5.21, because the joints are even closer to the point of rotation than the muscles are.

5.25 The formula for a straight line is $y = mx + b$, where m is the slope, and b is the y-intercept. Writing Equation 5.12, so that we get it in a similar form gives: $F/A = \gamma(\Delta L/L)$, so the slope is γ, which is Young's modulus.

5.28 Since vinegar expands much more than glass, without air in the bottle, the vinegar doesn't have any room to expand, except by breaking the glass. With air in the bottle, since air is highly compressible, the air will be compressed, giving the vinegar room to expand without breaking the bottle.

PROBLEMS

5.1 (a) To calculate the torque use Equation 5.4, where the perpendicular distance is 0.850 m, the force is 55.0 N, and the hinges are the pivot point.

$$\tau = r_\perp F = 0.850 \text{ m} \times 55.0 \text{ N} = 46.75 \text{ N} \cdot \text{m} = \boxed{46.8 \text{ N} \cdot \text{m}}$$

(b) It doesn't matter at what height you push. The torque depends on only the magnitude of the force applied and the perpendicular distance of the force's application from the hinges. (Children don't have a tougher time opening a door because they push *lower* than adults, they have a tougher time because they don't push far enough from the hinges.)

5.7 There are four forces acting on the horse and rider: **N** (acting straight up from the ground), **w** (acting straight down from the center of mass), *f* (acting horizontally to the left, at the ground to prevent the horse from slipping), and **F**$_{\text{wall}}$ (acting to the right). Since nothing is moving, the two conditions for equilibrium apply:

$$\text{net } F = 0 \text{ and net } \tau = 0.$$

The first condition leads to two equations (one for each direction):

$$\text{net } F_x = F_{\text{wall}} - f = 0 \text{ and net } F_y = N - w = 0$$

The torque equation (taking the torque about the center of gravity, where CCW is positive) gives:

$$\text{net } \tau = F_{\text{wall}} (1.40 \text{ m} - 1.20 \text{ m}) - f (1.40 \text{ m}) + N (0.350 \text{ m}) = 0.$$

The first two equations give:

$$F_{\text{wall}} = f, \text{ and } N = w = mg.$$

Substituting into the third equation gives:

$$F_{\text{wall}} (1.40 \text{ m} - 1.20 \text{ m}) - F_{\text{wall}} (1.40 \text{ m}) = -mg (0.350 \text{ m}),$$

so that the force on the wall is:

$$F_{\text{wall}} = \frac{mg (0.350 \text{ m})}{1.20 \text{ m}} = \frac{(500 \text{ kg})(9.80 \text{ m/s}^2)(0.350 \text{ m})}{1.20 \text{ m}} = 1429 \text{ N} = \boxed{1.43 \times 10^3 \text{ N}}$$

5.13 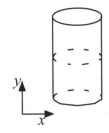 Use Equation 5.7 to calculate the center of mass. We know from symmetry that $x_{\text{CM}} = 0$ for all cases, and

$$y_{\text{CM}} = \frac{m_c y_{\text{CM(c)}} + m_s y_{\text{CM(s)}}}{m_c + m_s}, \text{ where } m_c = 0.0172 \text{ kg}, y_{\text{CM(c)}} = 0.00625 \text{ m}$$

(a) Half full means: $m_s = \dfrac{0.362 \text{ kg}}{2} = 0.181 \text{ kg}$ and $y_{\text{CM(s)}} = \dfrac{0.0125 \text{ m}}{4} = 0.003125 \text{ m}$, so:

$$x_{\text{CM}} = 0 \text{ m}, y_{\text{CM}} = \frac{(0.0172 \text{ kg})(0.00625 \text{ m}) + (0.181 \text{ kg})(0.003125 \text{ m})}{0.0172 \text{ kg} + 0.181 \text{ kg}} = \boxed{3.40 \text{ cm above base}}$$

(b) One tenth full means: $m_s = \dfrac{0.362 \text{ kg}}{10} = 0.0362 \text{ kg}$ and $y_{\text{CM(s)}} = \dfrac{0.0125 \text{ m}}{20} = 0.000625 \text{ m}$, so that

$$x_{\text{CM}} = 0 \text{ m}, y_{\text{CM}} = \frac{(0.0172 \text{ kg})(0.00625 \text{ m}) + (0.0362 \text{ kg})(0.000625 \text{ m})}{0.0172 \text{ kg} + 0.0362 \text{ kg}} = \boxed{2.44 \text{ cm above base}}$$

5.19   Looking at Figure 5.39, there are three forces acting on the entire sandwich board system: **w**, acting down at the center of mass of the system, and N_L and N_R, acting up at the ground for EACH of the legs. The tension and the hinge exert internal forces, and therefore cancel when considering the entire sandwich board. Using the first condition for equilibrium gives:

$$\text{net } F = N_L + N_R - w_s.$$

The normal forces are equal, due to symmetry, and the mass is given, so we can determine the normal forces:

$$2N = mg = (8.00 \text{ kg})(9.80 \text{ m/s}^2) \Rightarrow N = 39.2 \text{ N}.$$

Now, we can determine the tension in the chain and the force due to the hinge by using one side of the sandwich board:

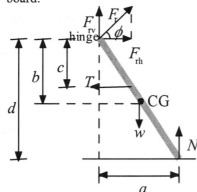

$$a = \frac{1.10 \text{ m}}{2} = 0.550 \text{ m}$$

$$b = \frac{1.30 \text{ m}}{2} = 0.650 \text{ m}$$

$$c = 0.500 \text{ m}$$

$$d = 1.30 \text{ m}$$

$$N = 39.2 \text{ N}$$

$$w = mg = \frac{8.00 \text{ kg}}{2}(9.80 \text{ m/s}^2) = 39.2 \text{ N} \quad (\text{for one side})$$

$$F_{rv} = F_r \sin\phi, F_{rh} = F_r \cos\phi$$

The system is in equilibrium, so the two conditions for equilibrium hold: net $F = 0$ and net $\tau = 0$
This gives three equations:

$$\text{net } F_x = F_{rh} - T = 0$$

$$\text{net } F_y = F_{rv} - w + N = 0$$

$$\text{net } \tau = -Tc - w\frac{a}{2} + Na = 0 \qquad\qquad (\text{with the pivot at hinge})$$

So that: $F_{rh} = T = F_r \cos\phi$, $F_{rv} = w - N = F_r \sin\phi$, and $Tc + w\frac{a}{2} = Na$.

(a) To solve for the tension, use the third equation: $Tc = Na - w\frac{a}{2}$. Since $N = w$, we have:

$$T = \frac{Na}{c} - \frac{wa}{2c} = \frac{wa}{2c} = \frac{(39.2 \text{ N})(0.550 \text{ m})}{2(0.500 \text{ m})} = \boxed{21.6 \text{ N}}$$

(b) To determine the force of the hinge, and the angle at which it acts, start with the second equation, remembering that $N = w$.

$$F_{rv} = w - N \Rightarrow F_{rv} = 0,$$

so either the force or the angle is zero. Now, the first equation says

$$F_{rh} = T,$$

so F_r cannot be zero, but rather $\phi = 0°$, giving a force of

$$\boxed{F_r = 21.6 \text{ N, acting horizontally}}$$

5.25   Use the definition of mechanical advantage, MA, from Equation 5.10:

$$\text{MA} = \frac{l_i}{l_o} = \frac{7.50 \text{ cm}}{0.480 \text{ cm}} = \boxed{15.6}$$

5.31 (a) Using Equations 5.9 and 5.10, we get an expression for the output force in terms of the input force and the two lever arms:

$$F_0 = F_i \times \frac{l_i}{l_0} = 50.0 \text{ N} \times \frac{11.0 \text{ cm}}{3.20 \text{ cm}} = \boxed{172 \text{ N}}$$

(b) Since the nut doesn't move, the total force on it must be $\boxed{0 \text{ N}}$. Friction keeps it from sliding out.

5.37 (a) Use the second condition for equilibrium: net $\tau = F_M (0.050 \text{ m}) - w(0.025 \text{ m}) = 0$, so that

$$F_M = w\frac{0.025 \text{ m}}{0.050 \text{ m}} = (50 \text{ N})\frac{0.025 \text{ m}}{0.050 \text{ m}} = \boxed{25 \text{ N}}$$

(b) To calculate the force on the joint, use the first condition for equilibrium: net $F_y = F_J - F_M - w = 0$, so that

$$F_J = F_M + w = (25 \text{ N}) + (50 \text{ N}) = \boxed{75 \text{ N}}$$

5.43 Use Equation 5.12, $\Delta L = \frac{1}{\gamma}\frac{F}{A}L_0$, where $\gamma = 16 \times 10^9$ N/m^2 (from Table 5.1), $L_0 = 0.350$ m,

$A = \pi r^2 = \pi(0.0180 \text{ m})^2 = 1.018 \times 10^{-3}$ m^2, and $F_{total} = 3w = 3(60.0 \text{ kg})(9.80 \text{ m/s}^2) = 1764$ N so that the force on each leg is $F_{leg} = F_{total}/2 = 882$ N. Substituting in the values gives:

$$\Delta L = \frac{1}{(16 \times 10^9 \text{ N/m}^2)}\frac{(882 \text{ N})}{(1.018 \times 10^{-3} \text{ m}^2)}(0.350 \text{ m}) = 1.90 \times 10^{-5} \text{ m}.$$

So each leg stretches by $\boxed{1.90 \times 10^{-3} \text{ cm}}$.

5.49 Use Equation 5.12, $\Delta L = \frac{1}{\gamma}\frac{F}{A}L_0$, where $L_0 = 6.00$ m, $\gamma = 210 \times 10^9$ N/m^2 (from Table 5.1). To calculate the

mass supported by the pipe, we need to add the mass of the new pipe to the mass of the 3.00 km piece of pipe and the mass of the drill bit: $m = m_p + m_{3 \text{ km}} + m_{bit} = m_p + (3.00 \times 10^3 \text{ m})(20.0 \text{ kg/m}) + 100 \text{ kg}$

$$m = (6.00 \text{ m})(20.0 \text{ kg/m}) + (3.00 \times 10^3 \text{ m})(20.0 \text{ kg/m}) + 100 \text{ kg} = 6.022 \times 10^4 \text{ kg},$$

so that the force on the pipe is:

$$F = w = mg = (6.022 \times 10^4 \text{ kg})(9.80 \text{ m/s}^2) = 5.902 \times 10^5 \text{ N}.$$

Finally, the cross sectional area is given by:

$$A = \pi r^2 = \pi\left(\frac{0.0500 \text{ m}}{2}\right)^2 = 1.963 \times 10^{-3} \text{ m}^2.$$

Substituting in the values gives:

$$\Delta L = \frac{1}{(210 \times 10^9 \text{ N/m}^2)}\frac{(5.902 \times 10^5 \text{ N})}{(1.963 \times 10^{-3} \text{ m}^2)}(6.00 \text{ m}) = 8.59 \times 10^{-3} \text{ m} = \boxed{8.59 \text{ mm}}$$

5.55   Given: $\dfrac{\Delta V}{V_0} = 2 \times 10^{-3}$ and $B = 1.8 \times 10^9 \text{ N/m}^2$. Find $\dfrac{F}{A}$. Using Equation 5.15, $\Delta V = \dfrac{1}{B}\dfrac{F}{A}V_0$, gives:

$$\frac{F}{A} = B\,\frac{\Delta V}{V_0} = \left(1.8 \times 10^9 \text{ N/m}^2\right)\left(2 \times 10^{-3}\right) = 3.6 \times 10^6 \text{ N/m}^2 = \boxed{4 \times 10^6 \text{ N/m}^2}\,.$$

Since $1 \text{ atm} = 1.013 \times 10^5 \text{ N/m}^2$, the pressure is about 36 atmospheres, far greater than the average jar is designed to withstand.

5.61   (a) Given: $\Delta L = 3 \times 10^{-5}$ m , $\gamma = 16 \times 10^9 \text{ N/m}^2$, $L_0 = 0.360$ m , and $r = 0.0175$ m so that

$A = \pi r^2 = \pi\left(0.0175 \text{ m}\right)^2 = 9.621 \times 10^{-4} \text{ m}^2$. Find: F. Using Equation 5.12 gives:

$$F = \gamma A\frac{\Delta L}{L_0} = \left(16 \times 10^9 \text{ N/m}^2\right)\left(9.621 \times 10^{-4} \text{ m}^2\right)\frac{3 \times 10^{-5} \text{ m}}{0.360 \text{ m}} = 1283 \text{ N} = \boxed{1 \times 10^3 \text{ N}}$$

(b) Use Newton's Second Law, Equation 4.1:

$$a = \frac{F}{m} = \frac{1283 \text{ N}}{62.0 \text{ kg}} = 20.69 \text{ m/s}^2 = \boxed{20.7 \text{ m/s}^2}$$

(c) Use Kinematics, Equation 2.11: $v^2 = v_0^2 + 2a\left(x - x_0\right)$, where $v_0 = 0$ m/s , and $x - x_0 = 0.100$ m .

$$v = \sqrt{v_0^2 + 2a\left(x - x_0\right)} = \sqrt{\left(0 \text{ m}^2/\text{s}^2\right) + 2\left(20.69 \text{ m/s}^2\right)\left(0.100 \text{ m}\right)} = \boxed{2.03 \text{ m/s}}$$

5.67   Given: $\gamma = 16 \times 10^9 \text{ N/m}^2$ (from Table 5.1), $L_0 = 0.140$ m , $d = 5.00 \times 10^{-3}$ m , $m_b = 3.11 \times 10^{-3}$ kg ,

$\Delta L = 38.0 \times 10^{-6}$ m , and $x - x_0 = 0.650$ m . Find v. Working backwards from what we want to find, we know that in order to calculate v, we first must know a. But, in order to calculate a, we must calculate F. So the first step is to use Equation 5.12 to calculate the force on the bullet. Next, using Newton's second law, we can calculate the acceleration, and finally, using kinematics we can calculate the final velocity (the muzzle velocity).

Using Equation 5.12, to get the force on the bullet:

$$F = \gamma A\frac{\Delta L}{L_0} = \left(16 \times 10^9 \text{ N/m}^2\right)\left[\pi\left(\frac{5.00 \times 10^{-3} \text{ m}}{2}\right)^2\right] \times \frac{38.0 \times 10^{-6} \text{ m}}{0.0140 \text{ m}} = 852.7 \text{ N} = 850 \text{ N}$$

Next, use Equation 4.1 to calculate the acceleration:

$$a = \frac{F}{m_b} = \frac{852.7 \text{ N}}{3.11 \times 10^{-3} \text{ kg}} = 2.742 \times 10^5 \text{ m/s}^2$$

Finally, use Equation 2.11: $v^2 = v_0^2 + 2a\left(x - x_0\right)$, where $v_0 = 0$ m/s , to calculate the final velocity:

$$v = \sqrt{v_0^2 + 2a\left(x - x_0\right)} = \sqrt{\left(0 \text{ m/s}\right)^2 + 2\left(2.742 \times 10^5 \text{ m/s}^2\right)\left(0.650 \text{ m}\right)} = 597 \text{ m/s} = \boxed{6.0 \times 10^2 \text{ m/s}}$$

6 WORK, ENERGY, AND POWER

CONCEPTUAL QUESTIONS

6.1 Holding a big box of books is something we think of as work in everyday circumstances. It is not, however, work in the scientific sense. The force that we exert is upward to hold the box, but there is no distance traveled, so there is no work. Energy is transferred from your muscles to the surroundings in the form of heat, but no work is done.

6.4 The lawn mower gains kinetic energy as its speed is increased. The mower loses kinetic energy as its speed is decreased.

6.7 The energy that the roller coaster has depends on the velocity squared, so the direction of the velocity does not affect the kinetic energy that it possesses. So, going up at 5 m/s instead of down at 5 m/s and ending at the same place gives the same change in kinetic energy and therefore the same final speed.

6.10 A closed system is a system that experiences only conserved forces. Energy can change form in it, but no energy can be lost in the conversion. Conserved forces are ones for which the work depends only on the starting and ending points, not on the path taken.

6.13 An open system is one that is subjected to non-conservative forces. For every non-conservative force that acts on the system, mechanical energy is either added or removed from the system. An open system may experience conservative forces that will not change the mechanical energy of the system.

6.16 No, devices with efficiencies less than one lose energy, but they do not violate the conservation of energy principle because that lost energy is converted into another type of energy, like heat for instance.

6.19 The spark does not injure you because it is only acting for a very short time. Therefore, the work done by the spark is small, even though the power is very large, so no damage is done.

6.22 When rubbing your hands together, energy is produced, which is why they feel warmer. The heat produced by rubbing your hands together is internal work generated on each hand, so no work is done on the outside world. Since no work is done on the outside world, the efficiency of rubbing your hands together is quite large.

PROBLEMS

6.1   Using Equation 6.1, where $F = 5.00$ N, $d = 0.600$ m, and since the force is applied horizontally, $\theta = 0°$:

$$W = Fd\cos\theta = (5.00 \text{ N})(0.600 \text{ m})\cos 0° = \boxed{3.00 \text{ J}}.$$

Using the conversion factor: 1 kcal = 4186 J, gives:

$$W = 3.00 \text{ J} \times \frac{1 \text{ kcal}}{4186 \text{ J}} = \boxed{7.17 \times 10^{-4} \text{ kcal}}.$$

6.7   (a) The work done by friction is in the opposite direction of the motion, so $\theta = 180°$, and therefore

$$W_f = Fd\cos\theta = 35.0 \text{ N} \times 20.0 \text{ m} \times \cos 180° = \boxed{-700 \text{ J}}$$

(b) The work done by gravity is perpendicular to the direction of motion, so $\theta = 90°$, and

$$W_g = Fd\cos\theta = 35.0 \text{ N} \times 20.0 \text{ m} \times \cos 90° = \boxed{0 \text{ J}}$$

(c) If the cart moves at a constant speed, no energy is transferred to it, from the work-energy theorem:

$$\text{net } W = W_s + W_f = 0, \text{ or } W_s = \boxed{700 \text{ J}}.$$

(d) Use Equation 6.1,

$$W_s = Fd\cos\theta,$$

where $\theta = 25°$, and solve for the force:

$$F = \frac{W_s}{d\cos\theta} = \frac{700 \text{ J}}{20.0 \text{ m} \times \cos 25°} = 38.62 \text{ N} = \boxed{38.6 \text{ N}}$$

(e) Since there is no change in speed, the work energy theorem says that there is no net work done on the cart:

$$\text{net } W = W_f + W_s = -700 \text{ J} + 700 \text{ J} = \boxed{0 \text{ J}}$$

6.13 Use the work energy theorem (Equations 6.1 and 6.2):

$$\text{net } W = \frac{1}{2}mv^2 - \frac{1}{2}mv_0^2 = Fd\cos\theta,$$

where $d = 0.200$ m, $m = 900$ kg, $v = 0$ m/s, $v_0 = 1.12$ m/s, and $\theta = 0°$. Solving for the force gives:

$$F = \frac{mv^2 - mv_0^2}{2d\cos\theta} = \frac{(900 \text{ kg})(0 \text{ m/s})^2 - (900 \text{ kg})(1.12 \text{ m/s})^2}{2(0.200 \text{ m})\cos 0°} = \boxed{-2.82 \times 10^3 \text{ N}}.$$

The force is negative because the car is decelerating.

6.19 (a) Using Equation 6.4, $\Delta PE_g = mgh$, where $m = 5.00 \times 10^{13}$ kg, $g = 9.80$ m/s^2, and $h = 40.0$ m, gives:

$$\Delta PE_g = (5.00 \times 10^{13} \text{ kg})(9.80 \text{ m/s}^2)(40.0 \text{ m}) = \boxed{1.96 \times 10^{16} \text{ J}}.$$

(b) From Table 6.1, we know the energy stored in a 9-megaton fusion bomb is 3.8×10^{16} J, so that

$$\frac{E_{\text{lake}}}{E_{\text{bomb}}} = \frac{1.96 \times 10^{16} \text{ J}}{3.8 \times 10^{16} \text{ J}} = \boxed{0.52}.$$

The energy of stored in the lake is approximately 1/2 that of a 9-megaton fusion bomb.

6.25 Use the work energy theorem (Equation 6.6):

$$\mathrm{KE_i + PE_i = KE_f + PE_f}, \text{ where } \mathrm{KE} = \frac{1}{2}mv^2, \text{ and } \mathrm{PE} = mgh.$$

(a) Given: $v_i = 5.00 \text{ m/s}$, $h_i = 25.0 \text{ m}$, and $h_f = 0 \text{ m}$. Find: v_f.

So: $\frac{1}{2}mv_i^2 + mgh_i = \frac{1}{2}mv_f^2 + \cancel{mgh_f}$, or:

$$v_f = \sqrt{2gh_i + v_i^2} = \sqrt{2(9.80 \text{ m/s}^2)(25 \text{ m}) + (5.00 \text{ m/s})^2} = 22.69 \text{ m/s} = \boxed{22.7 \text{ m/s}}$$

(b) Taking the results of part (a) to be our initial conditions: $v_i = 22.69 \text{ m/s}$, $h_i = 0 \text{ m}$, and $h_f = 25.0 \text{ m} - 20.0 \text{ m} = 5.0 \text{ m}$. Find v_f.

Conserving energy gives: $\frac{1}{2}mv_i^2 + \cancel{mgh_i} = \frac{1}{2}mv_f^2 + mgh_f$, so:

$$v_f = \sqrt{v_i^2 - 2gh_f} = \sqrt{(22.69 \text{ m/s})^2 - 2(9.80 \text{ m/s}^2)(5.0 \text{ m})} = 20.41 \text{ m/s} = \boxed{20.4 \text{ m/s}}$$

(c) Taking the starting point as the initial conditions: $v_i = 5.00 \text{ m/s}$, $v_f = 10.0 \text{ m/s}$, and $h_i = 0 \text{ m}$ (because we want the height below the starting point). Find: $-h_f$.

Conserving energy gives: $\frac{1}{2}mv_i^2 + \cancel{mgh_i} = \frac{1}{2}mv_f^2 + mgh_f$, so:

$$h_f = \frac{v_i^2 - v_f^2}{2g} = \frac{(5.00 \text{ m/s})^2 - (10.0 \text{ m/s})^2}{2(9.80 \text{ m/s}^2)} = -3.83 \text{ m}.$$

So the roller coaster is $\boxed{3.83 \text{ m}}$ below the starting point.

6.31 From Table 6.2: $P_{\text{Crab}} = 10^{28} \text{ W}$, and $P_{\text{supernova}} = 5 \times 10^{37} \text{ W}$, so that

$$\frac{P}{P_0} \approx \frac{10^{28} \text{ W}}{5 \times 10^{37} \text{ W}} = \boxed{2 \times 10^{-10}}.$$

The power today is 10 orders of magnitude smaller than it was at the time of the explosion.

6.37 (a) Use Equation 6.9, (where t is in seconds!)

$$P = \frac{W}{t} = \frac{6.00 \times 10^6 \text{ J}}{(8.00 \text{ h})(3600 \text{ s/1 h})} = 208.3 \text{ J/s} = \boxed{208 \text{ W}}$$

(b) Use the work energy theorem to express the work needed to lift the bricks: $W = mgh$, where $m = 2000 \text{ kg}$ and $h = 1.50 \text{ m}$. Then use Equation 6.9 to solve for the time:

$$P = \frac{W}{t} = \frac{mgh}{t} \Rightarrow t = \frac{mgh}{P} = \frac{(2000 \text{ kg})(9.80 \text{ m/s}^2)(1.50 \text{ m})}{208.3 \text{ W}} = 141.1 \text{ s} = \boxed{141 \text{ s}}$$

6.43

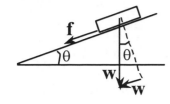

The energy supplied by the engine is converted into frictional energy as the car goes up the incline. Use Equations 6.1 and 6.9 to get:

$$P = \frac{W}{t} = \frac{Fd}{t} = F\left(\frac{d}{t}\right) = Fv \text{, where } F \text{ is parallel to the incline and must oppose}$$

the resistive forces and the force of gravity: $F = f + w = 600 \text{ N} + mg\sin\theta$, where $f = 600 \text{ N}$, $m = 950 \text{ kg}$, $\theta = 2.00°$, and $v = 30.0 \text{ m/s}$, so that:

$P = (f + mg\sin\theta)v$, or:

$$P = \left[600 \text{ N} + (950 \text{ kg})(9.80 \text{ m/s}^2)\sin 2°\right](30.0 \text{ m/s}) = \boxed{2.77 \times 10^4 \text{ W}}$$

6.49 Use the work energy theorem to determine the work done by the shot-putter:

$$\text{net } W = \frac{1}{2}mv^2 + mgh - \frac{1}{2}\cancel{mv_0^2} - \cancel{mgh_0} = \frac{1}{2}(7.27 \text{ kg})(14.0 \text{ m/s})^2 + (7.27 \text{ kg})(9.80 \text{ m/s}^2)(0.800 \text{ m}) = 769.5 \text{ J}$$

The power can be found using Equation 6.9:

$$P = \frac{W}{t} = \frac{769.5 \text{ J}}{1.20 \text{ s}} = 641.2 \text{ W} = \boxed{641 \text{ W}} .$$

Then, using the conversion 1 hp = 746 W , we see that

$$P = 641.2 \text{ W} \times \frac{1 \text{ hp}}{746 \text{ W}} = 0.8595 \text{ hp} = \boxed{0.860 \text{ hp}} .$$

6.55

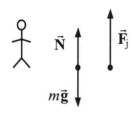

(a) Given: $m = 80.0 \text{ kg}$, $h = 0.600 \text{ m}$, and $d = 0.0150 \text{ m}$. Find: net F. Using Equation 6.1 and the work-energy theorem gives: $W = F_j d = mgh$. Thus, the force in the joint is:

$$F_j = \frac{mgh}{d} = \frac{(80.0 \text{ kg})(9.80 \text{ m/s}^2)(0.600 \text{ m})}{0.0150 \text{ m}} = 3.136 \times 10^4 \text{ N} .$$

Looking at the free body diagram: net $F = w + F_j$, so that:

$$\text{net } F = (80.0 \text{ kg})(9.80 \text{ m/s}^2) + 3.136 \times 10^4 \text{ N} = \boxed{3.21 \times 10^4 \text{ N}}$$

(b) Now, let $d = 0.300 \text{ m}$, so that $F_j = \frac{mgh}{d} = \frac{(80.0 \text{ kg})(9.80 \text{ m/s}^2)(0.600 \text{ m})}{0.300 \text{ m}} = 1568 \text{ N}$ and

$$\text{net } F = (80.0 \text{ kg})(9.80 \text{ m/s}^2) + 1568 \text{ N} = \boxed{2.35 \times 10^3 \text{ N}}$$

(c) In (a), $\dfrac{\text{net } F}{mg} = \dfrac{32,144 \text{ N}}{784 \text{ N}} = \boxed{41.0}$ (This could be damaging to the body!)

In (b), $\dfrac{\text{net } F}{mg} = \dfrac{2352 \text{ N}}{784 \text{ N}} = \boxed{3.00}$ (This can be easily sustained.)

6.61 (a) To calculate the potential energy use Equation 6.4, where $m = 7 \times 10^9$ kg and $h = \frac{1}{4} \times 146$ m $= 36.5$ m :

$$PE = mgh = \left(7.00 \times 10^9 \text{ kg}\right)\left(9.80 \text{ m/s}^2\right)\left(36.5 \text{ m}\right) = 2.504 \times 10^{12} \text{ J} = \boxed{2.50 \times 10^{12} \text{ J}}$$

(b) First, we need to calculate the energy needed to feed the 1000 workers over the 20 years:

$$E_{\text{in}} = NPt = 1000 \times \frac{300 \text{ kcal}}{\text{h}} \times \frac{4186 \text{ J}}{\text{kcal}} \times 20 \text{ y} \times \frac{330 \text{ d}}{\text{y}} \times \frac{12 \text{ h}}{\text{d}} = 9.946 \times 10^{13} \text{ kcal} .$$

Now, since the workers must provide the PE from part (a), use Equation 6.10a to calculate their efficiency:

$$Eff = \frac{W_{\text{out}}}{E_{\text{in}}} = \frac{PE}{E_{\text{in}}} = \frac{2.504 \times 10^{12} \text{ J}}{9.946 \times 10^{13} \text{ J}} = 0.0252 = \boxed{2.52\%}$$

(c) If each worker requires 3600 kcal/day, and we know the composition of their diet, we can calculate the mass of the food required: $E_{\text{protein}} = \left(3600 \text{ kcal}\right)\left(0.05\right) = 180 \text{ kcal}$; $E_{\text{carbohydrate}} = \left(3600 \text{ kcal}\right)\left(0.60\right) = 2160 \text{ kcal}$; and $E_{\text{fat}} = \left(3600 \text{ kcal}\right)\left(0.35\right) = 1260 \text{ kcal}$. Now, from Table 6.1 we can convert the energy required into the mass required for each component of their diet:

$$m_{\text{protein}} = E_{\text{protein}} \times \frac{1 \text{ g}}{4.1 \text{ kcal}} = 180 \text{ kcal} \times \frac{1 \text{ g}}{4.1 \text{ kcal}} = 43.90 \text{ g} ;$$

$$m_{\text{carbohydrate}} = E_{\text{carbohydrate}} \times \frac{1 \text{ g}}{4.1 \text{ kcal}} = 2160 \text{ kcal} \times \frac{1 \text{ g}}{4.1 \text{ kcal}} = 526.8 \text{ g} ; \text{ and}$$

$$m_{\text{fat}} = E_{\text{fat}} \times \frac{1 \text{ g}}{9.3 \text{ kcal}} = 1260 \text{ kcal} \times \frac{1 \text{ g}}{9.3 \text{ kcal}} = 135.5 \text{ g} .$$

Therefore, the total mass of food required for the average worker per day is:

$$m_{\text{person}} = m_{\text{protein}} + m_{\text{carbohydrate}} + m_{\text{fat}} = \left(43.90 \text{ g}\right) + \left(526.8 \text{ g}\right) + \left(135.5 \text{ g}\right) = 706.2 \text{ g} ,$$

and the total amount of food required for the 20,000 workers is:

$$m = Nm_{\text{person}} = 20,000 \times 0.7062 \text{ kg} = 1.41 \times 10^4 \text{ kg} = \boxed{1.4 \times 10^4 \text{ kg}}$$

6.67

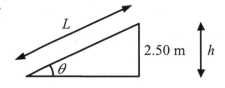

Given: $m = 60.0$ kg , $v_0 = 12.0$ m/s , $h = 2.5$ m , $\theta = 35°$, $\mu_k = 0.0800$. Find: v . First, use trig to find the distance traveled along the incline: $\sin \theta = \frac{h}{L} \Rightarrow L = \frac{h}{\sin \theta} = \frac{2.50 \text{ m}}{\sin 35°} = 4.359$ m

There are three forces acting: **w** acting straight down, f acting down the incline, and **N** acting perpendicular to the incline. Using Newton's Laws gives:

$$\text{net } F = -mg \sin \theta - f = ma , \text{ and}$$

$$\text{net } F_\perp = N - mg \cos \theta ,$$

so that $f = \mu_k N = \mu_k mg \cos \theta$. Solving for a gives

$$a = \frac{-\mu_k mg \cos \theta - mg \sin \theta}{m} = -g\left(\mu_k \cos \theta + \sin \theta\right) = -\left(9.80 \text{ m/s}^2\right)\left[\left(0.0800\right)\cos 35° + \sin 35°\right] = -6.263 \text{ m/s}^2 .$$

Now, using the kinematics Equation 2.11: $v^2 = v_0^2 + 2aL$, we can find v:

$$v = \sqrt{\left(12.0 \text{ m/s}\right)^2 + 2\left(-6.263 \text{ m/s}^2\right)\left(4.359 \text{ m}\right)} = \boxed{9.46 \text{ m/s}}$$

This problem is lengthier than using energy considerations because it requires two steps rather than one.

LINEAR MOMENTUM

CONCEPTUAL QUESTIONS

7.1 The small mass will have the larger kinetic energy because it has the larger velocity, and the kinetic energy is proportional to the mass times the velocity squared, while the momentum is only proportional to the mass times the velocity.

7.4 A small force acting over a long time can impart the same momentum as a large force acting over a short time.

7.7 Momentum can indeed be conserved if there are external forces acting, as long as there is no NET external force.

7.10 Objects in a system can indeed have momentum while the system has zero momentum. This is true when the internal momentums add to be zero. For instance, a system containing two children on ice can have zero momentum if the two children start out at rest, holding onto each other. Then, if the children push away from each other, they will each have momentum that is non-zero, but the system of the two children will have zero momentum by conservation of momentum.

7.13 An inelastic collision does not conserve internal kinetic energy, although it will conserve momentum if the net external force is zero. A perfectly inelastic collision occurs when two objects collide and stick together.

7.16 The momentum of the center of mass is unchanged by the explosion. If the pieces experienced significant air resistance, the center of mass would slow relative to the unexploded shell.

7.19 (a) If we ignore the rotation of the cube and assume that it is much more massive than the small mass, the angle at which the small mass will bounce off is independent of the angle b. The small mass will always reflect off at twice the angle the cube face makes with the horizontal.

(b) For a sphere, a small impact parameter will lead to a large angle, θ_1, near $180°$. If b is large, so the mass only collides at a glancing angle, the angle, θ_1, will be small, and can be near $0°$.

The reason there is a difference between parts (a) and (b) is because the angle of approach by the small mass varies with b for the case of the sphere, while it is fixed for the case of the cube.

PROBLEMS

7.1 (a) The momentum is calculated using Equation 7.1:
$$p_e = m_e v_e = 2000 \text{ kg} \times 7.50 \text{ m/s} = \boxed{1.50 \times 10^4 \text{ kg} \cdot \text{m/s}}$$

(b) $p_b = m_b v_b = 0.0400 \text{ kg} \times 600 \text{ m/s} = 24.0 \text{ kg} \cdot \text{m/s}$, so
$$\frac{p_e}{p_b} = \frac{1.50 \times 10^4 \text{ kg} \cdot \text{m/s}}{24.0 \text{ kg} \cdot \text{m/s}} = \boxed{625}.$$
The momentum of the elephant is much larger because the mass of the elephant is much larger.

(c) $p_h = m_h v_h = 90.0 \text{ kg} \times 7.40 \text{ m/s} = \boxed{6.66 \times 10^2 \text{ kg} \cdot \text{m/s}}$.
Again, the momentum is smaller than that of the elephant because the mass of the hunter is much smaller.

7.7   (a) Use Equation 7.2 to calculate the net force exerted on the hand:

$$\text{net } F = \frac{\Delta p}{\Delta t} = \frac{m\Delta v}{\Delta t} = \frac{1.50 \text{ kg}\left(0 \text{ m/s} - 4.00 \text{ m/s}\right)}{2.50 \times 10^{-3} \text{ s}} = -2.40 \times 10^{3} \text{ N}$$

(taking toward the leg as the positive direction). Therefore, by Newton's third law, the net force exerted on the leg is $\boxed{2.40 \times 10^{3} \text{ N, toward the leg}}$

(b) The force on each hand would have the same magnitude as that found in part (a) (but in opposite directions by Newton's third law) because the change in momentum and time interval are the same.

7.13   Given: $m = 1.00 \times 10^{7} \text{ kg}$, $v_{0} = 0.750 \text{ m/s}$, $v = 0 \text{ m/s}$ $\Delta x = 6.00 \text{ m}$. Find: net F on the pier. First, we need a way to express the time, Δt, in terms of knowns. Using Equations 2.7 and 2.8 gives: $\Delta x = \bar{v}\Delta t = \frac{1}{2}\left(v + v_{0}\right)\Delta t$ so that

$$\Delta t = \frac{2\Delta x}{v + v_{0}} = \frac{2\left(6.00 \text{ m}\right)}{\left(0 + 0.750\right) \text{ m/s}} = 16.0 \text{ s} .$$

Now, use Equation 7.2 to calculate the net force on the ship:

$$\text{net } F = \frac{\Delta p}{\Delta t} = \frac{m\left(v - v_{0}\right)}{\Delta t} = \frac{\left(1.00 \times 10^{7} \text{ kg}\right)\left(0 - 0.750\right) \text{ m/s}}{16.0 \text{ s}} = -4.69 \times 10^{5} \text{ N} .$$

So, by Newton's third law, the net force on the pier is $\boxed{4.69 \times 10^{5} \text{ N}}$, in the original direction of the ship.

7.19   Use conservation of momentum, Equation 7.4: $m_{1}v_{1} + m_{2}v_{2} = m_{1}v' + m_{2}v'$, since their final velocities are the same.

$$v' = \frac{m_{1}v_{1} + m_{2}v_{2}}{m_{1} + m_{2}} = \frac{\left(150,000 \text{ kg}\right)\left(0.300 \text{ m/s}\right) + \left(110,000 \text{ kg}\right)\left(-0.120 \text{ m/s}\right)}{150,000 \text{ kg} + 110,000 \text{ kg}} = \boxed{0.122 \text{ m/s}} .$$

The final velocity is in the direction of the first car because it had a larger initial momentum.

7.25   (a) Use conservation of momentum to find the mass of the barbell: $m_{1}v_{1} + m_{2}v_{2} = m_{1}v_{1}' + m_{2}v_{2}'$, where $v_{1} = v_{2} = 0 \text{ m/s}$, and $v_{1}' = -0.500 \text{ m/s}$ (since it recoils backwards), so solving for the mass of the barbell gives:

$$0 = m_{1}v_{1} + m_{2}v_{2} \Rightarrow m_{2} = \frac{-m_{1}v_{1}}{v_{2}} = \frac{-\left(80.0 \text{ kg}\right)\left(-0.500 \text{ m/s}\right)}{10.0 \text{ m/s}} = \boxed{4.00 \text{ kg}}$$

(b) Find the change in kinetic energy: $\Delta KE = \frac{1}{2}m_{1}v_{1}'^{2} + \frac{1}{2}m_{2}v_{2}'^{2} - \frac{1}{2}m_{1}v_{1}^{2} - \frac{1}{2}m_{2}v_{2}^{2} = \frac{1}{2}\left(m_{1}v_{1}'^{2} + m_{2}v_{2}'^{2}\right)$

$$\Delta KE = \frac{1}{2}\left[\left(80.0 \text{ kg}\right)\left(-0.500 \text{ m/s}\right)^{2} + \left(4.00 \text{ kg}\right)\left(10.0 \text{ m/s}\right)^{2}\right] = \boxed{210 \text{ J}}$$

(c) The clown does work to throw the barbell, so the kinetic energy comes from the muscles of the clown. The muscles convert the chemical potential energy of ATP into kinetic energy.

7.31 (a) Use conservation of momentum for the player and the ball: $m_1v_1 + m_2v_2 = (m_1 + m_2)v'$ so that

$$v' = \frac{m_1v_1 + m_2v_2}{m_1 + m_2} = \frac{(110 \text{ kg})(8.00 \text{ m/s}) + (0.410 \text{ kg})(25.0 \text{ m/s})}{110 \text{ kg} + 0.410 \text{ kg}} = 8.063 \text{ m/s} = \boxed{8.06 \text{ m/s}}.$$

(b) Find the change in kinetic energy: $\Delta KE = KE' - (KE_1 + KE_2) = \frac{1}{2}(m_1 + m_2)v'^2 - \frac{1}{2}(m_1v_1^2 + m_2v_2^2)$

$$\Delta KE = \frac{1}{2}(110.41 \text{ kg})(8.063 \text{ m/s})^2 - \frac{1}{2}\left[(110 \text{ kg})(8.00 \text{ m/s})^2 + (0.410 \text{ kg})(25.0 \text{ m/s})^2\right] = \boxed{-59.0 \text{ J}}$$

(c) (i) $v' = \dfrac{(110 \text{ kg})(8.00 \text{ m/s}) + (0.410 \text{ kg})(-25.0 \text{ m/s})}{110.41 \text{ kg}} = 7.877 \text{ m/s} = \boxed{7.88 \text{ m/s}}$ (in direction of player)

(ii) $\Delta KE = \frac{1}{2}(110.41 \text{ kg})(7.877 \text{ m/s})^2 - \frac{1}{2}\left[(110 \text{ kg})(8.00 \text{ m/s})^2 + (0.410 \text{ kg})(-25.0 \text{ m/s})^2\right] = \boxed{-223 \text{ J}}$

7.37 (a) Given: $v_1 = v_2 = 0 \text{ m/s}$, $m_1 = 3.00 \text{ kg}$. Use conservation of momentum: $m_1v_1 + m_2v_2 = m_1v_1' + m_2v_2'$

$$m_1v_1' = -m_2v_2' \Rightarrow v_1' = \frac{-m_2v_2'}{m_1} = \frac{-(0.0250 \text{ kg})(550 \text{ m/s})}{3.00 \text{ kg}} = -4.583 \text{ m/s} = \boxed{-4.58 \text{ m/s}}$$

(b) The rifle begins at rest, so $KE_i = 0 \text{ J}$, and $\Delta KE = \frac{1}{2}m_1v_1'^2 = \frac{1}{2}(3.00 \text{ kg})(-4.58 \text{ m/s})^2 = \boxed{31.5 \text{ J}}$

(c) Now, $m_1 = 28.0 \text{ kg}$, so that $v_1' = \frac{-m_2v_2}{m_1} = \frac{-(0.0250 \text{ kg})(550 \text{ m/s})}{28.0 \text{ kg}} = \boxed{-0.491 \text{ m/s}}$

(d) Again, $KE_i = 0 \text{ J}$, and $\Delta KE = \frac{1}{2}m_1v_1'^2 = \frac{1}{2}(28.0 \text{ kg})(-0.491 \text{ m/s})^2 = 3.376 \text{ J} = \boxed{3.38 \text{ J}}$

(e) Conceptual Question 7.1 makes the observation that if two objects have the same momentum, the heavier object will have a smaller kinetic energy. Keeping the rifle close to the body, increases the effective mass of the rifle, hence reducing the kinetic energy of the recoiling rifle. Since pain is related to the amount of kinetic energy, a rifle hurts less if it is held against the body.

7.43 (a) Use the definition of KE to calculate the initial speed of the He atom:

$$\frac{1}{2}m_1v_1^2 = KE_i \Rightarrow v_1 = \left(\frac{2KE_i}{m_1}\right)^{1/2} = \left[\frac{2(8.00 \times 10^{-13} \text{ J})}{6.68 \times 10^{-27} \text{ kg}}\right]^{1/2} = 1.548 \times 10^7 \text{ m/s}$$

Conservation of internal kinetic energy gives: $\frac{1}{2}m_1v_1^2 = \frac{1}{2}m_1v_1'^2 + \frac{1}{2}m_2v_2'^2$, or $\qquad \frac{m_1}{m_2}(v_1^2 - v_1'^2) = v_2'^2$ (1)

Conservation of momentum along the x-axis gives: $\qquad m_1v_1 = m_1v_1'\cos\theta_1 + m_2v_2'\cos\theta_2$ (2)

Conservation of momentum along the y-axis gives: $\qquad 0 = m_1v_1'\sin\theta_1 + m_2v_2'\sin\theta_2$ (3)

Rearranging Equations (2) and (3) gives: $\qquad m_1v_1 - m_1v_1'\cos\theta_1 = m_2v_2'\cos\theta_2$ (2′)

$$-m_1v_1'\sin\theta_1 = m_2v_2'\sin\theta_2 \quad (3')$$

Squaring Equations (2′) and (3′) and adding gives:

$$m_2^2 v_2'^2 \cos^2 \theta_2 + m_2^2 v_2'^2 \sin^2 \theta_2 = (m_1 v_1 - m_1 v_1' \cos \theta_1)^2 + (-m_1 v_1' \sin \theta_1)^2,$$

or

$$m_2^2 v_2'^2 = m_1^2 v_1^2 - 2m_1^2 v_1 v_1' \cos \theta_1 + m_1^2 v_1'^2 \qquad (4)$$

Solving (4) for $v_2'^2$ and substituting into (1) gives:

$$\frac{m_1}{m_2}\left(v_1^2 - v_1'^2\right) = \frac{m_1^2}{m_2^2}\left(v_1^2 + v_1'^2 - 2v_1 v_1' \cos \theta_1\right)$$

or

$$v_1^2 - v_1'^2 = \frac{m_1}{m_2}\left(v_1^2 + v_1'^2 - 2v_1 v_1' \cos \theta_1\right) \text{ so that } \left(1 + \frac{m_1}{m_2}\right)v_1'^2 - \left(\frac{2m_1}{m_2}v_1 \cos \theta_1\right)v_1' - \left(1 - \frac{m_1}{m_2}\right)v_1^2 = 0$$

Substituting in the values: $v_1 = 1.548 \times 10^7$ m/s; $\theta_1 = 120°$; $m_1 = 6.68 \times 10^{-27}$ kg; $m_2 = 3.29 \times 10^{-25}$ kg, gives:

$$a \equiv 1 + \frac{m_1}{m_2} = 1.0203,$$

$$b \equiv -\frac{2m_1}{m_2}v_1 \cos \theta_1 = 3.143 \times 10^5 \text{ m/s, and}$$

$$c \equiv -\left(1 - \frac{m_1}{m_2}\right)v_1^2 = -2.348 \times 10^{14} \text{ m}^2/\text{s}^2$$

So that

$$v_1' = \frac{-b \pm \sqrt{b^2 - 4ac}}{2a} = \frac{-3.143 \times 10^5 \text{ m/s} + \sqrt{\left(3.143 \times 10^5 \text{ m/s}\right)^2 - 4(1.0203)\left(-2.348 \times 10^{14} \text{ m}^2/\text{s}^2\right)}}{2(1.0203)}$$

or

$$v_1' = \boxed{1.50 \times 10^7 \text{ m/s}} \text{ and } v_2' = \sqrt{\frac{m_1}{m_2}\left(v_1^2 - v_1'^2\right)} = \boxed{5.36 \times 10^5 \text{ m/s}}$$

Finally, solving for the angle gives:

$$\tan \theta_2 = \frac{-v_1' \sin \theta_1}{v_1 - v_1' \cos \theta_1} = \frac{-\left(1.50 \times 10^7 \text{ m/s}\right)\sin 120°}{1.548 \times 10^7 \text{ m/s} - \left(1.50 \times 10^7 \text{ m/s}\right)\cos 120°} = -0.56529$$

or:

$$\theta_2 = \tan^{-1}(-0.56529) = \boxed{-29.5°}$$

(b) The final kinetic energy is then: $KE_f = (0.5)m_1 v_1'^2 = (0.5)\left(6.68 \times 10^{-27} \text{ kg}\right)\left(1.50 \times 10^7 \text{ m/s}\right)^2 = \boxed{7.52 \times 10^{-13} \text{ J}}$

7.49 Use Equation 7.8: $v = v_0 + v_e \ln\left(\frac{m_0}{m}\right)$, where $m_0 = 4000$ kg, $m = 4000$ kg $- 3500$ kg $= 500$ kg, and

$v_e = 2.00 \times 10^3$ m/s, so that

$$v - v_0 = \left(2.00 \times 10^3 \text{ m/s}\right)\ln\left(\frac{4000 \text{ kg}}{500 \text{ kg}}\right) = 4.159 \times 10^3 \text{ m/s} = \boxed{4.16 \times 10^3 \text{ m/s}}.$$

7.55 (a) Given: $m_{total} = 5.00$ kg, $v = 0$ m/s, $m_2 = 0.250$ kg, $m_1 = m_{total} - m_2 = (5.00 \text{ kg}) - (0.250 \text{ kg}) = 4.75$ kg, $v_2' = 10.0$ m/s. Find: $v_{1,f}'$. First, find v_1', the velocity after ejecting the fluid, using conservation of momentum:

$$(m_1 + m_2)v = m_1 v_1' + m_2 v_2', \text{ so that } v_1' = \frac{-m_2 v_2'}{m_1} = \frac{-(0.250 \text{ kg})(10.0 \text{ m/s})}{4.75 \text{ kg}} = -0.526 \text{ m/s}.$$

Now, the frictional force slows the squid over the 0.100 s, so using Equation 7.2:
$\Delta p = (\text{net } F)t = m_1 v_{1,f}' + m_2 v_2'$, gives:

$$v_{1,f}' = \frac{ft - m_2 v_2'}{m_1} = \frac{(5.00 \text{ N})(0.100 \text{ s}) - (0.250 \text{ kg})(10.0 \text{ m/s})}{4.75 \text{ kg}} = -0.421 \text{ m/s}.$$

Therefore, the squid recoils at $\boxed{0.421 \text{ m/s}}$

(b) The change in kinetic energy of the squid is: $\Delta KE = \frac{1}{2} m_1 v_{1,f}'^2 - \frac{1}{2} m_1 v_1'^2 = \frac{1}{2} m_1 \left(v_{1,f}'^2 - v_1'^2 \right)$ so that

$$\Delta KE = \frac{1}{2}(4.75 \text{ kg})\left[(-0.421 \text{ m/s})^2 - (-0.526 \text{ m/s})^2 \right] = \boxed{-0.236 \text{ J}}$$

7.61 (a) Given: $R = 200$ m, and $\theta = 45°$. Find v_0. Use Equation 3.13: $R = \frac{v_0^2 \sin 2\theta}{g}$ to get:

$$v_0 = \sqrt{\frac{gR}{\sin 2\theta}} = \sqrt{\frac{(9.80 \text{ m/s}^2)(200 \text{ m})}{\sin 90°}} = 44.27 \text{ m/s} = \boxed{44.3 \text{ m/s}}$$

(b) Use Equation 7.1: $\Delta p = p_f - p_i = m(v_f - v_i) = (0.0450 \text{ kg})(44.27 \text{ m/s} - 0 \text{ m/s}) = \boxed{1.99 \text{ kg} \cdot \text{m/s}}$

(c) Use Equation 7.2: $\text{net } F = \frac{\Delta p}{\Delta t} = \frac{1.992 \text{ kg} \cdot \text{m/s}}{1.00 \times 10^{-3} \text{ s}} = 1992 \text{ N} = \boxed{1.99 \times 10^3 \text{ N}}$

(d) Use Equation 5.14: $\Delta x = \frac{1}{S} \frac{F}{A} L_0 = \frac{1}{80 \times 10^9 \text{ N/m}^2} \frac{1992 \text{ N}}{\pi (0.00500 \text{ m})^2} (0.800 \text{ m}) = \boxed{2.5 \times 10^{-4} \text{ m}}$

7.67 (a) Given: $m_1 = 4.00$ kg, $v_1 = 3.00$ m/s, $v_2 = 0$ m/s, $v' = 3.50$ m/s. Find: m_2. Using conservation of momentum: $m_1 v_1 = (m_1 + m_2)v_1'$, so that:

$$m_2 = \frac{m_1 v_1}{v_1'} - m_1 = m_1 \left(\frac{v_1}{v_1'} - 1 \right) = 4.00 \text{ kg} \left(\frac{3.00 \text{ m/s}}{3.50 \text{ m/s}} - 1 \right) = \boxed{-0.571 \text{ kg}}$$

(b) The mass has a negative value. Masses must be positive!

(c) The final speed of the combination must always be less than the initial speed of the incoming projectile!

8 UNIFORM CIRCULAR MOTION AND GRAVITATION

CONCEPTUAL QUESTIONS

8.1   The rotation angle is analogous to distance and the angular velocity is analogous to velocity.

8.4   Centripetal force is any force causing uniform circular motion. Theoretically, any force can cause uniform circular motion and therefore be a centripetal force, however, circular motion requires the force to act perpendicular to the velocity and therefore friction alone cannot cause circular motion. Combinations of forces can be centripetal forces if they act perpendicular to the velocity and cause uniform circular motion.

8.7   The radius of curvature of Path 2 is larger than that of Path 1, so the driver can actually have a larger speed and still experience the same centripetal force. This allows the car to travel faster, without losing control around the corner.

8.10   The lunch box will follow path B, and the trail will follow a path curved toward the center of the merry-go-round, ending at the outer edge of the merry-go-round.

8.13   The mass wants to continue in a STRAIGHT line path, so since the mass is being kept in a circle, the string must pull it toward the center of the circle at all times. The string feels an equal and opposite force (by Newton's third law) to the force it exerts on the mass, causing there to be a very real force stretching the string.

8.16   The force that pins the riders to the wall is actually the tendency for the rider to want to travel in a straight line. It is the normal force of the barrel pushing on the person that keeps the rider traveling in a circle rather than a straight line. The other forces acting on the person are gravity downward and friction upward. If the barrel is spinning fast enough, the frictional force is large enough to hold the person up. The frictional force is proportional to the normal force keeping the rider traveling in a circle, which depends on the speed of the rotation by Equation 8.7.

8.19

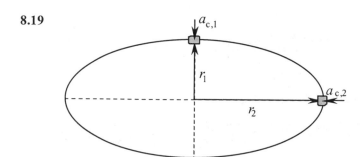

The satellite travels faster when it is closer to the parent body because the force of gravity is greater, leading to a larger centripetal acceleration and also a larger speed.

In other words: $a_c = \dfrac{v^2}{r} = \dfrac{GM}{r^2} \Rightarrow v = \sqrt{\dfrac{GM}{r}}$,

so that when $r_1 < r_2$, $v_1 > v_2$.

PROBLEMS

8.1   Given: $d = 1.15 \text{ m} \Rightarrow r = \dfrac{1.15 \text{ m}}{2} = 0.575 \text{ m}$, $\Delta\theta = 200{,}000 \text{ rot} \times \dfrac{2\pi \text{ rad}}{1 \text{ rot}} = 1.257 \times 10^6 \text{ rad}$. Find: Δs. Use

Equation 8.1: $\Delta\theta = \dfrac{\Delta s}{r}$, so that

$$\Delta s = \Delta\theta \times r = \left(1.257 \times 10^6 \text{ rad}\right)\left(0.575 \text{ m}\right) = 7.226 \times 10^5 \text{ m} = \boxed{723 \text{ km}}$$

8.7   Given: $r = 0.420 \text{ m}$, $v = 32.0 \text{ m/s}$. Find: ω. Use Equation 8.4: $\omega = \dfrac{v}{r} = \dfrac{32.0 \text{ m/s}}{0.420 \text{ m}} = \boxed{76.2 \text{ rad/s}}$. Finally,

convert to rpm by using the conversion factor: $1 \text{ rev} = 2\pi \text{ rad}$,

$$\omega = 76.2 \text{ rad/s} \times \dfrac{1 \text{ rev}}{2\pi \text{ rad}} \times \dfrac{60 \text{ s}}{1 \text{ min}} = 728 \text{ rev/min} = \boxed{728 \text{ rpm}}.$$

8.13   (a)   Use Equation 8.4 to find the linear velocity:

$$v = r\omega = \left(0.100 \text{ m}\right)\left(50{,}000 \text{ rev/min} \times \dfrac{2\pi \text{ rad}}{1 \text{ rev}} \times \dfrac{1 \text{ min}}{60 \text{ s}}\right) = 524 \text{ m/s} = \boxed{0.524 \text{ km/s}}$$

   (b)   Given: $\omega = 2\pi \text{ rad/y} \times \dfrac{1 \text{ y}}{3.16 \times 10^7 \text{ s}} = 1.988 \times 10^{-7} \text{ rad/s}$, $r = 1.496 \times 10^{11} \text{ m}$. Use Equation 8.4 to find the

linear velocity:

$$v = r\omega = \left(1.496 \times 10^{11} \text{ m}\right)\left(1.988 \times 10^{-7} \text{ rad/s}\right) = 2.975 \times 10^4 \text{ m/s} = \boxed{29.7 \text{ km/s}}$$

8.19   Using the equation on page 200 gives:

$$\tan\theta = \dfrac{v^2}{rg} \Rightarrow v = \sqrt{rg \tan\theta} = \sqrt{\left(100 \text{ m}\right)\left(9.80 \text{ m/s}^2\right)\tan 20.0°} = \boxed{18.9 \text{ m/s}}$$

8.25   (a)   Using the equation on page 204 gives:

$$g = \dfrac{GM}{r^2} \Rightarrow M = \dfrac{r^2 g}{G} = \dfrac{\left(6371 \times 10^3 \text{ m}\right)^2 \left(9.830 \text{ m/s}^2\right)}{6.673 \times 10^{-11} \text{ N} \cdot \text{m}^2 / \text{kg}^2} = \boxed{5.979 \times 10^{24} \text{ kg}}$$

   (b)   This is identical to the best value to three significant figures.

8.31   (a)   Use Equation 8.8 to calculate the force:

$$F_f = \dfrac{GMm}{r^2} = \dfrac{\left(6.673 \times 10^{-11} \text{ N} \cdot \text{m}^2 / \text{kg}^2\right)\left(100 \text{ kg}\right)\left(4.20 \text{ kg}\right)}{\left(0.200 \text{ m}\right)^2} = \boxed{7.01 \times 10^{-7} \text{ N}}$$

   (b)   The mass of Jupiter is: $m_J = 1.90 \times 10^{27} \text{ kg}$, so using Equation 8.8 gives:

$$F_J = \dfrac{\left(6.673 \times 10^{-11} \text{ N} \cdot \text{m}^2 / \text{kg}^2\right)\left(1.90 \times 10^{27} \text{ kg}\right)\left(4.20 \text{ kg}\right)}{\left(6.29 \times 10^{11} \text{ m}\right)^2} = \boxed{1.35 \times 10^{-6} \text{ N}}. \quad \dfrac{F_f}{F_J} = \dfrac{7.01 \times 10^{-7} \text{ N}}{1.35 \times 10^{-6} \text{ N}} = \boxed{0.521}$$

8.37   Using Equation 8.11: $\dfrac{r^3}{T^2} = \dfrac{G}{4\pi^2}M$, we can solve for the mass of Jupiter:

$$M_J = \frac{4\pi^2}{G} \times \frac{r^3}{T^2} = \frac{4\pi^2}{6.673\times10^{-11}\ \text{N}\cdot\text{m}^2/\text{kg}^2} \times \frac{\left(4.22\times10^8\ \text{m}\right)^3}{\left[(0.00485\ \text{y})\left(3.16\times10^7\ \text{s/y}\right)\right]^2} = \boxed{1.89\times10^{27}\ \text{kg}}$$

8.43   (a) Use Equation 8.6: $F_c = ma_c$ substituting using Equations 8.7 and 8.8: $\dfrac{GmM}{r^2} = \dfrac{mv^2}{r}$, we have:

$$v = \sqrt{\frac{GM_E}{r_s}} = \sqrt{\frac{\left(6.673\times10^{-11}\ \text{N}\cdot\text{m}^2/\text{kg}^2\right)\left(5.979\times10^{24}\ \text{kg}\right)}{900\times10^3\ \text{m}}} = \boxed{2.11\times10^4\ \text{m/s}}$$

(b)

In the satellite's frame of reference, the rivet has two perpendicular velocity components equal to v from part (a):

$$v_{tot} = \sqrt{v^2 + v^2} = \sqrt{2}v = \sqrt{2}\left(2.105\times10^4\ \text{m/s}\right) = \boxed{2.98\times10^4\ \text{m/s}}$$

(c) Using kinematics:

$$d = v_{tot}t \Rightarrow t = \frac{d}{v_{tot}} = \frac{3.00\times10^{-3}\ \text{m}}{2.98\times10^4\ \text{m/s}} = \boxed{1.01\times10^{-7}\ \text{s}}$$

(d) Using Equation 7.2:

$$\overline{F} = \frac{\Delta p}{\Delta t} = \frac{mv_{tot}}{t} = \frac{\left(0.500\times10^{-3}\ \text{kg}\right)\left(2.98\times10^4\ \text{m/s}\right)}{1.01\times10^{-7}\ \text{s}} = \boxed{1.48\times10^8\ \text{N}}$$

(e) The energy is generated from the rivet. In the satellite's frame of reference, $v_i = v_{tot}$, and $v_f = 0$. So the change in kinetic energy of the rivet is:

$$\Delta\text{KE} = \frac{1}{2}mv_{tot}^2 - \frac{1}{2}mv_i^2 = \frac{1}{2}\left(0.500\times10^{-3}\ \text{kg}\right)\left(2.98\times10^4\ \text{m/s}\right)^2 - 0\ \text{J} = \boxed{2.22\times10^5\ \text{J}}$$

ROTATIONAL MOTION AND ANGULAR MOMENTUM

CONCEPTUAL QUESTIONS

9.1 Angular acceleration is analogous to linear acceleration. Torque is analogous to force. Moment of inertia is analogous to mass. Work is still work. Rotational kinetic energy is analogous to translational kinetic energy. Angular momentum is analogous to linear momentum. The change in angular momentum is analogous to impulse.

9.4 (a) When the record starts to spin, the bug experiences both tangential and centripetal acceleration.

(b) When the record rotates at a constant angular velocity, the bug experiences centripetal acceleration.

(c) When the record slows, the bug again experiences both tangential and centripetal acceleration.

9.7 Since $\tau = rF\sin\theta$, a small force exerted on a crowbar produces a large torque on a lug nut because the lever arm is quite long. When you stand out of the way and hold a door open for someone to pass, you exert a large force on the door, but create a small torque because you apply your force very close to the hinge of the door. Little kids have this problem when they don't understand to push the door on the side without the hinges.

9.10 When the yo-yo is thrown downward, there is initially potential energy because of its starting height and kinetic energy because of its being thrown down. When it reaches the bottom, it has converted this energy into rotational energy. When it climbs back up, it converts the rotational energy back into potential energy and kinetic energy while keeping some of its rotational energy.

9.13 The angular momentum of the car must be conserved at the exact time you rev the engine, so since the engine rotates in one direction, the car must rock in the opposite direction. Angular momentum is only conserved initially, since the car loses energy to heat and therefore will not conserve angular momentum forever.

9.16 The helicopter rotates in the direction opposite to the main lifting blades in order to conserve angular momentum. Or, in terms of Newton's Third Law, for every action there is an equal and opposite reaction, the reaction of the helicopter is in the opposite direction to the action of the main lifting blades.

9.19 The global heating causes the atmosphere to expand, and in order to conserve angular momentum; the earth must slow its rotation. This is similar to when the skater in Example 9.13 has her arms outstretched. Since a day is measured as the time it takes the earth to rotate once on its axis, the slower rotation means that the length of a day is longer.

9.22 The force on the astronaut is a reaction force from his turning the bolt, so he rotates in the opposite direction to the bolt. The satellite, however, feels a frictional force from the bolt, and so rotates in the direction the bolt rotates. If the astronaut holds onto the satellite with a handhold, he can prevent the counter-rotation because he would add a force on the satellite in the direction of his motion, or counter to the direction of the bolt, thereby canceling the rotation caused by the bolt.

9.25 The horse's large muscles far from the hip means that the lever arm for the torque the muscle creates is large, therefore allowing the muscle to produce a large torque. The torque is proportional to the angular acceleration and the moment of inertia of the leg, so since the moment of inertia of the leg is kept small by keeping the distribution of the mass of the leg nearer the hip, the angular accelerations are quite large and the horse can run quite fast.

9.28 The motorcycle must conserve angular momentum, so rotating the handlebar to the right will cause the rest of the motorcycle to rotate slightly to the left. This is the same principle you are taught in driving school about turning INTO a skid.

PROBLEMS

9.1 First, convert the speed to m/s:

$$v = \frac{500 \text{ km}}{1 \text{ h}} \times \frac{1 \text{ h}}{3600 \text{ s}} \times \frac{1000 \text{ m}}{1 \text{ km}} = 138.9 \text{ m/s}.$$

Then, use Equation 8.4 to determine the angular speed:

$$\omega = \frac{v}{r} = \frac{138.9 \text{ m/s}}{30.0 \text{ m}} = 4.630 \text{ rad/s}.$$

Finally, convert the angular speed to rev/s:

$$\omega = 4.630 \text{ rad/s} \times \frac{1 \text{ rev}}{2\pi \text{ rad}} = \boxed{0.737 \text{ rev/s}}$$

9.7 (a) Using the results of Example 9.7: $\alpha = 4.44 \text{ rad/s}^2$, $\omega_0 = 0.00 \text{ rad/s}$, and $\omega = 1.50 \text{ rad/s}$, we can solve for time using Equation 9.5b,

$$\omega = \omega_0 + \alpha t, \text{ or } t = \frac{\omega - \omega_0}{\alpha} = \frac{(1.50 \text{ rad/s}) - (0 \text{ rad/s})}{4.44 \text{ rad/s}^2} = \boxed{0.338 \text{ s}}.$$

(b) Now, to find θ without using our result from part (a), use Equation 9.5d: $\omega^2 = \omega_0^2 + 2\alpha\theta$, giving:

$$\theta = \frac{\omega^2 - \omega_0^2}{2\alpha} = \frac{(1.50 \text{ rad/s})^2 - (0 \text{ rad/s})^2}{2(4.44 \text{ rad/s}^2)} = 0.253 \text{ rad} \times \frac{1 \text{ rev}}{2\pi \text{ rad}} = \boxed{0.0403 \text{ rev}}.$$

Note: you can also use Equation 9.5a or Equation 9.5c, and you will get the same answer.

Generally, it is safer to use an equation that does not require your answer to previous parts, in case you made a mistake. If you have time in an exam, you can always solve these types of problems using two different equations and they should give the same answer (this allows you to check your work).

(c) Given: $F = -300 \text{ N}$, $r = 1.35 \text{ m}$, $I = 84.38 \text{ kg} \cdot \text{m}^2$, $\omega = 0 \text{ rad/s}$, and $\omega_0 = 1.50 \text{ rad/s}$. Find: t. To get an expression for the angular acceleration, use Equation 9.7:

$$\alpha = \frac{\text{net } \tau}{I} = \frac{rF}{I}.$$

Then, to find the time, use Equation 9.5b:

$$t = \frac{\omega - \omega_0}{\alpha} = \frac{(\omega - \omega_0)I}{rF} = \frac{(0 \text{ rad/s} - 1.50 \text{ rad/s})(84.38 \text{ kg} \cdot \text{m}^2)}{(1.35 \text{ m})(-300 \text{ N})} = \boxed{0.313 \text{ s}}$$

9.13

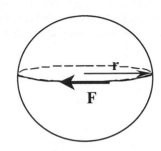

Step 1: There is a torque present due to a force being applied perpendicular to a rotation axis. The mass involved is the earth.

Step 2: The system of interest is the earth.

Step 3: The free body diagram is drawn to the left.

Step 4: Given: $F = -4.00 \times 10^7$ N , $r = r_E = 6.376 \times 10^6$ m , $M = 5.979 \times 10^{24}$ kg ,

$$\omega_0 = \frac{1 \text{ rev}}{24.0 \text{ h}} \times \frac{2\pi \text{ rad}}{1 \text{ rev}} \times \frac{1 \text{ h}}{3600 \text{ s}} = 7.272 \times 10^{-5} \text{ rad/s} \text{, and}$$

$$\omega = \frac{1 \text{ rev}}{28.0 \text{ h}} \times \frac{2\pi \text{ rad}}{1 \text{ rev}} \times \frac{1 \text{ h}}{3600 \text{ s}} = 6.233 \times 10^{-5} \text{ rad/s} \text{. Find: } t.$$

Use Equation 9.7 to determine the angular acceleration:

$$\alpha = \frac{\text{net } \tau}{I} = \frac{rF}{2Mr^2/5} = \frac{5F}{2Mr}$$

Now, that we have an expression for the angular acceleration, we can use Equation 9.5b to get the time:

$$\omega = \omega_0 + \alpha t \Rightarrow t = \frac{\omega - \omega_0}{\alpha} = \frac{(\omega - \omega_0)2Mr}{5F}. \text{ Substituting in the numbers gives:}$$

$$t = \frac{2(6.233 \times 10^{-5} \text{ rad/s} - 7.272 \times 10^{-5} \text{ rad/s})(5.979 \times 10^{24} \text{ kg})(6.376 \times 10^6 \text{ m})}{5(-4.00 \times 10^7 \text{ N})} = \boxed{3.97 \times 10^{18} \text{ s}}$$

Superman doesn't have to worry about Zorch for about 126 billion years!!

9.19 The moment of inertia for the wheel is

$$I = \frac{M}{2}\left(R_1^2 + R_2^2\right) = \frac{12.0 \text{ kg}}{2}\left[(0.280 \text{ m})^2 + (0.330 \text{ m})^2\right] = 1.124 \text{ kg} \cdot \text{m}^2.$$

Using Equation 9.10:

$$KE_{rot} = \frac{1}{2}I\omega^2 = \frac{1}{2}(1.124 \text{ kg} \cdot \text{m}^2)(120 \text{ rad/s})^2 = \boxed{8.09 \times 10^3 \text{ J}}$$

9.25 (a) Assuming her arm starts extended vertically downward, we can calculate the initial angular acceleration.

Given: $\theta = 60° \times \dfrac{2\pi \text{ rad}}{360°} = 1.047 \text{ rad}$, $m_w = 2.00$ kg , $r_w = 0.240$ m , $I = 0.250 \text{ kg} \cdot \text{m}^2$, and $F_M = 750$ N ,

where $r_\perp = 0.0200$ m . Find: α .

The only force that contributes to the torque when the mass is vertical is the muscle, and the moment of inertia is that of the arm and that of the mass, so using Equation 9.7, we can find the angular acceleration:

$$\alpha = \frac{\text{net } \tau}{I + m_w r_w^2} = \frac{F_M r_\perp}{I + m_w r_w^2}, \text{ so that}$$

$$\alpha = \frac{(750 \text{ N})(0.0200 \text{ m})}{0.250 \text{ kg} \cdot \text{m}^2 + (2.00 \text{ kg})(0.240 \text{ m})^2} = 41.07 \text{ rad/s}^2 = \boxed{41.1 \text{ rad/s}^2}$$

(b) The work done can be calculated by Equation 9.8:

$$\text{net } W = (\text{net } \tau)\theta = F_M r_\perp \theta = (750 \text{ N})(0.0200 \text{ m})(1.047 \text{ rad}) = \boxed{15.7 \text{ J}}$$

9.31 (a) The moment of inertia for the earth around the sun is $I = MR^2$, since the earth is like a point object. Using Equation 9.11, we can calculate the angular momentum of the earth around the sun:

$$L_{orb} = I\omega = MR^2\omega = \left(5.979\times10^{24}\text{ kg}\right)\left(1.496\times10^{11}\text{ m}\right)^2\left(\frac{2\pi\text{ rad}}{3.16\times10^7\text{ s}}\right) = \boxed{2.66\times10^{40}\text{ kg}\cdot\text{m}^2/\text{s}}$$

(b) The moment of inertia for the earth on its axis is $I = \dfrac{2MR^2}{5}$, since the earth is a solid sphere. We can calculate the angular momentum of the earth on its axis, by using Equation 9.11:

$$L_{rot} = I\omega = \left(\frac{2}{5}MR^2\right)\omega = \frac{2}{5}\left(5.979\times10^{24}\text{ kg}\right)\left(6.376\times10^6\text{ m}\right)^2\left(\frac{2\pi\text{ rad}}{24\times3600\text{ s}}\right) = \boxed{7.07\times10^{33}\text{ kg}\cdot\text{m}^2/\text{s}}.$$

The angular momentum of the earth in its orbit around the sun is 3.76×10^6 times larger than the angular momentum of the earth around its axis.

9.37 (a) The final moment of inertia is again the disk plus the stick, but this time, the radius for the disk is smaller:

$$I' = mr^2 + \frac{MR^2}{3} = \left(0.0500\text{ kg}\right)\left(0.100\text{ m}\right)^2 + \left(0.667\text{ kg}\right)\left(1.20\text{ m}\right)^2 = 0.961\text{ kg}\cdot\text{m}^2.$$

The final angular velocity can then be determined following the solution to part (a) of Example 9.15:

$$\omega' = \frac{mvr}{I'} = \frac{(0.0500\text{ kg})(30.0\text{ m/s})(0.100\text{ m})}{0.961\text{ kg}\cdot\text{m}^2} = \boxed{0.156\text{ rad/s}}$$

(b) The kinetic energy before the collision is the same as in Example 9.15: $KE = \boxed{22.5\text{ J}}$. The final kinetic energy is now: $KE' = \dfrac{1}{2}I'\omega'^2 = \dfrac{1}{2}\left(0.961\text{ kg}\cdot\text{m}^2\right)\left(0.156\text{ rad/s}\right)^2 = \boxed{1.17\times10^{-2}\text{ J}}$

(c) The initial linear momentum is the same as in Example 9.15: $p = \boxed{1.50\text{ kg}\cdot\text{m/s}}$. The final linear momentum is then $p' = mr\omega' + \dfrac{M}{2}R\omega' = \left[mr + \left(\dfrac{M}{2}\right)R\right]\omega'$, so that:

$$p' = \left[\left(0.0500\text{ kg}\right)\left(0.100\text{ m}\right) + \left(1.00\text{ kg}\right)\left(1.20\text{ m}\right)\right]\left(0.156\text{ rad/s}\right) = \boxed{0.188\text{ kg}\cdot\text{m/s}}$$

9.43 (a) Given: $M = 90.0\text{ kg}$, $R = 0.340\text{ m}$ (for the solid disk), $\omega = 90.0\text{ rot/min}$, $N = 20.0\text{ N}$, and $\mu_k = 0.20$. Find: α. The frictional force is given by: $f = \mu_k N = 0.20(20.0\text{ N}) = 4.0\text{ N}$. *This frictional force is reducing the speed of the grindstone, so the angular acceleration will be negative.* Using the moment of inertia for a solid disk and Equation 9.7, we know: $\tau = -fR = I\alpha = \dfrac{1}{2}MR^2\alpha$. Solving for the angular acceleration gives:

$$\alpha = \frac{-2f}{MR} = \frac{-2(4.0\text{ N})}{(90.0\text{ kg})(0.340\text{ m})} = -0.261\text{ rad/s}^2 = \boxed{-0.26\text{ rad/s}^2}\text{ (2 sig. figs due to }\mu_k\text{)}.$$

(b) Given: $\omega = 0\text{ rad/s}$, $\omega_0 = \dfrac{90.0\text{ rev}}{\text{min}}\times\dfrac{2\pi\text{ rad}}{\text{rev}}\times\dfrac{1\text{ min}}{60\text{ s}} = 9.425\text{ rad/s}$. Find: θ. Using rotational kinematics Equation 9.5d: $\omega^2 - \omega_0^2 = 2\alpha\theta$, so that:

$$\theta = \frac{\omega^2 - \omega_0^2}{2\alpha} = \frac{(0\text{ rad/s})^2 - (9.425\text{ rad/s})^2}{2(-0.261\text{ rad/s}^2)} = 169.9\text{ rad}\times\frac{1\text{ rev}}{2\pi\text{ rad}} = 27.0\text{ rev} = \boxed{27\text{ rev}}.$$

FLUID STATICS $\quad$ 10

CONCEPTUAL QUESTIONS

10.1 A fluid is a state of matter that yields to sideways or shearing forces; solids do not. A fluid flows, while a solid retains its shape without the aid of a container.

10.4 The density of air is greater closer to the earth. It decreases rapidly, and approximately linearly, with altitude.

10.7 A dull hypodermic needle has more surface area in contact with the skin as it is entering, so it hurts more because the force on the skin is proportional to the surface area of contact.

10.10 There is more pressure on the toes when toe dancing because only the tips of the toes are in contact with the ground and must support the entire force of gravity on the body. The larger the surface area in contact with the ground, the less pressure exerted.

10.13 Your body has grown accustomed to supporting the pressure of atmosphere. The difference in pressure between sunbathing and standing is rather small, and is easily overcome by our bodies.

10.16 This will produce a larger force at the slave than if the master cylinder and the slave are at the same height because the fluid at the master cylinder will start at a higher potential energy and can therefore do more work.

10.19 Atmospheric pressure will not add to the pressure in a rigid tank, but it will add to the pressure in a toy balloon. In general, atmospheric pressure will not affect the total pressure in a fluid when an incompressible container surrounds the fluid.

10.22 The height of an open manometer containing mercury is approximately 100 mm, which is quite easy to have on a doctor's wall. The height necessary for a manometer containing water would be approximately 1360 mm, which becomes cumbersome to have in a doctor's office, and especially if it needs to be portable. Since mercury has a larger density than water, the height produced is smaller given the same pressure applied.

10.25 More force is required to pull the plug in a full bathtub because the water exerts pressure downward on the plug and that pressure must be overcome to pull the plug. This does not contradict Archimedes' principle because there is no water underneath the plug before it is pulled.

10.28 No, the water will not overflow when the ice melts because the density of ice is smaller than the density of water, so the ice takes up more space as ice than it will when it melts.

10.31 Surface tension is due to cohesive force between molecules which cause the surface of a liquid to act like a stretched rubber sheet.

10.34 Water forms beads on an oily surface because the cohesive forces responsible for surface tension are larger than the adhesive forces, which tend to flatten the drop. On skin that is not oiled, water beads are flattened because the adhesive forces between water and skin are strong, overcoming surface tension.

10.37 Surface tension in the alveoli creates a positive pressure opposing inhalation. You can exhale without muscle action by letting surface tension in the alveoli create its positive pressure.

PROBLEMS

10.1   From Table 10.1: $\rho_{Au} = 19.32$ g/cm^3, so using Equation 10.1, $\rho = \dfrac{m}{V}$, we have:

$$V = \frac{m}{\rho} = \frac{31.103 \text{ g}}{19.32 \text{ g/cm}^3} = \boxed{1.610 \text{ cm}^3}$$

10.7   (a) From Table 10.1: $\rho_{gas} = 0.680 \times 10^3$ kg/m^3, so using Equation 10.1: $\rho = \dfrac{m}{V} = \dfrac{m}{lwh}$, so the height is:

$$h = \frac{m}{\rho l w} = \frac{50.0 \text{ kg}}{\left(0.680 \times 10^3 \text{ kg/m}^3\right)\left(0.900 \text{ m}\right)\left(0.500 \text{ m}\right)} = \boxed{0.163 \text{ m}}$$

   (b) The volume of this gasoline tank is 19.4 gallons, quite reasonably sized for a passenger car.

10.13 Using Equation 10.2, we can solve for the pressure:

$$P = \frac{F}{A} = \frac{mg}{\pi r^2} = \frac{\left(1.00 \times 10^{-3} \text{ kg}\right)\left(9.80 \text{ m/s}^2\right)}{\pi \left(2.00 \times 10^{-4} \text{ m}\right)^2} = \boxed{7.80 \times 10^4 \text{ Pa}} .$$

This pressure is approximately 585 mm Hg.

10.19 (a) Using Equation 10.2, we can solve for the pressure:

$$P = \frac{F}{A} = \frac{0.300 \text{ N}}{1.10 \text{ cm}^2} \times \left(\frac{100 \text{ cm}}{1 \text{ m}}\right)^2 = 2.73 \times 10^3 \text{ Pa} \times \frac{1 \text{ mm Hg}}{133.3 \text{ Pa}} = \boxed{20.5 \text{ mm Hg}}$$

   (b) From Table 10.5, we see that the range of pressures in the eye is 12-24 mm Hg, so the result in part (a) is within that range.

10.25

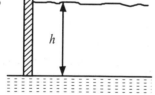

The average pressure on a dam is given by Equation 10.3: $\bar{P} = (h/2)\rho g$, where $h/2$ is the average height of the water behind the dam. Then, the force on the dam is found using Equation 10.2: $P = F/A$, so that

$$F = \bar{P}A = \left(\frac{h}{2}\rho g\right)(hL), \text{ or } F = \frac{\rho g h^2 L}{2} .$$

Thus, the average force on a rectangular dam increases with the square of the depth.

10.31 (a) Using Equation 10.5, $\dfrac{F_1}{A_1} = \dfrac{F_2}{A_2}$, we see that the ratio of the areas becomes: $\dfrac{A_S}{A_M} = \dfrac{F_S}{F_M} = \dfrac{100}{1} = \boxed{100}$

(b) We know that the area goes as: $\pi r^2 = \pi \dfrac{d^2}{4}$, so the ratio of the areas gives:

$$\frac{A_S}{A_M} = \frac{\pi r_S^2}{\pi r_M^2} = \frac{\pi (d_S/2)^2}{\pi (d_M/2)^2} = \frac{d_S^2}{d_M^2} = 100 \text{ , so that } \frac{d_S}{d_M} = \sqrt{100} = \boxed{10.0}$$

(c) Since the work input equals the work output, and work is proportional to force times distance,

$$F_i d_i = F_o d_o \Rightarrow \frac{d_o}{d_i} = \frac{F_i}{F_o} = \frac{1}{100} .$$

This tells us that the distance through which the output force moves is reduced by a factor of $\boxed{100}$, relative to the distance through which the input force moves.

10.37 (a) If the input cylinder is moved a distance d_i , it displaces a volume of fluid V, where the volume of fluid displaced must be the same for the input as the output:

$$V = d_i A_i = d_o A_o \Rightarrow d_o = d_i \left(\frac{A_i}{A_o} \right) .$$

Now, using Equation 10.5, we can write the ratio of the areas in terms of the ratio of the forces:

$$\frac{F_1}{A_1} = \frac{F_2}{A_2} \Rightarrow F_o = F_i \left(\frac{A_o}{A_i} \right) .$$

Finally, writing the work output in terms of force and distance gives:

$$W_o = F_o d_o = \left(\frac{F_i A_o}{A_i} \right) \left(\frac{d_i A_i}{A_o} \right) = F_i d_i = W_i .$$

In other words, the work output equals the work input for a hydraulic system.

(b) If the system is not moving, the friction would not play a role. With friction, we know there are losses, so that $W_{out} = W_{in} - W_f$, therefore, the work output is less than the work input. In other words, with friction, you need to push harder on the input piston than was calculated. Note: the volume of fluid is still conserved.

10.43 (a) The apparent mass loss is equal to the mass of the fluid displaced, so the mass of the fluid displaced is just the difference between the mass of the bone and its apparent mass:

$$m_{\text{displaced}} = 45.0 \text{ g} - 3.60 \text{ g} = \boxed{41.4 \text{ g}}$$

(b) Using Archimedes' principle, we know that the volume of water displaced equals the volume of the bone, so using Equation 10.1 we see that

$$V_b = V_w = \frac{m_w}{\rho_w} = \frac{41.4 \text{ g}}{1.00 \text{ g/cm}^3} = \boxed{41.4 \text{ cm}^3}$$

(c) Using Equation 10.1, we can calculate the average density of the bone:

$$\bar{\rho}_b = \frac{m_b}{V_b} = \frac{45.0 \text{ g}}{41.4 \text{ cm}^3} = \boxed{1.09 \text{ g/cm}^3} .$$

This is clearly not the density of the bone everywhere. The air pockets will have a density of approximately 1.29×10^{-3} g/cm^3 , while the bone will be substantially denser.

10.49 (a) From Equation 10.8, fraction submerged $= \dfrac{\overline{\rho}_{obj}}{\rho_{fl}}$, we see that:

$$\overline{\rho}_{person} = \rho_{fresh\ water} \times (\text{fraction submerged}) = \left(1.00 \times 10^3\ \text{kg/m}^3\right)(0.960) = \boxed{960\ \text{kg/m}^3}$$

(b) The density of seawater is greater than that of fresh water, so she should float more.

$$\text{fraction submerged} = \frac{\overline{\rho}_{person}}{\rho_{seawater}} = \frac{960\ \text{kg/m}^3}{1025\ \text{kg/m}^3} = 0.9366 \ . \quad \text{Therefore, the percent of her volume above water is:}$$

$$\% \text{ above water} = (1.0000 - 0.9366) \times 100\% = \boxed{6.34\%}\ .$$

She does indeed float more in seawater.

10.55 To determine if the ingot is gold or tungsten, we need to calculate the percent difference between the two substances both out and in water. Then, the difference between these percent differences is the necessary accuracy that we must have in order to determine the substance we have. The percent difference is calculated by calculating the difference in a quantity and dividing that by the value for gold.
Out of water: Using the difference in density, the percent difference is then:

$$\%_{out} = \frac{\rho_g - \rho_t}{\rho_g} \times 100\% = \frac{19.32\ \text{g/cm}^3 - 19.30\ \text{g/cm}^3}{19.32\ \text{g/cm}^3} \times 100\% = \boxed{0.1035\% \text{ in air}}$$

In water: Assume a 1.000 cm^3 nugget. Then the apparent mass loss is equal to that of the water displaced, i.e., 1.000 g. So, we can calculate the percent difference in the mass loss by using the difference in masses:

$$\%_{in} = \frac{m'_g - m'_t}{m'_g} \times 100\% = \frac{18.32\ \text{g} - 18.30\ \text{g}}{18.32\ \text{g}} \times 100\% = \boxed{0.1092\% \text{ in water}}$$

The difference between the required accuracies for the two methods is

$$\text{difference} = 0.1092\% - 0.1035\% = \boxed{5.7 \times 10^{-3}\ \%}\ ,$$

so we need 5 digits of accuracy to determine the difference between gold and tungsten.

10.61 Use Equation 10.12 to find the height to which capillary action will move sap through the xylem tube:

$$h = \frac{2\gamma \cos\theta}{\rho g r} = \frac{2(0.0728\ \text{N/m})(\cos 0°)}{\left(1050\ \text{kg/m}^3\right)\left(9.80\ \text{m/s}^2\right)\left(2.50 \times 10^{-5}\ \text{m}\right)} = \boxed{0.566\ \text{m}}$$

10.67 (a) Use Equation 10.11 to find the gauge pressure inside a spherical soap bubble of radius 1.50 cm:

$$P_1 = \frac{4\gamma}{r} = \frac{4(0.0370\ \text{N/m})}{(0.0150\ \text{m})} = \boxed{9.87\ \text{N/m}^2}$$

(b) Use Equation 10.11 to find the gauge pressure inside a spherical soap bubble of radius 4.00 cm:

$$P_2 = \frac{4\gamma}{r} = \frac{4(0.0370\ \text{N/m})}{0.0400\ \text{m}} = \boxed{3.70\ \text{N/m}^2}$$

(c) If they form one bubble without losing any air, then the total volume remains constant,

$$V = V_1 + V_2 = \frac{4}{3}\pi r_1^3 + \frac{4}{3}\pi r_2^3 = \frac{4}{3}\pi R^3 \ .$$

Solving for the single bubble radius gives:

$$R = \left[r_1^3 + r_2^3 \right]^{1/3} = \left[(0.0150\ \text{m})^3 + (0.0400\ \text{m})^3 \right]^{1/3} = 0.0406\ \text{m} \ .$$

So we can calculate the gauge pressure for the single bubble using Equation 10.11:

Chapter 10: Fluid Statics

$$P = \frac{4\gamma}{R} = \frac{4(0.0370 \text{ N/m})}{0.0406 \text{ m}} = \boxed{3.65 \text{ N/m}^2}$$

10.73 The negative gauge pressure that can be achieved is the sum of the pressure due to the water pressure and the pressure in the lungs:

$$P = -3.00 \text{ cm H}_2\text{O} - 60.0 \text{ cm H}_2\text{O} = \boxed{-63.0 \text{ cm H}_2\text{O}}$$

10.79 (a) This part is a unit conversion problem:

$$P_0 = (10.0 \text{ mm Hg})\left(\frac{133 \text{ N/m}^2}{1.0 \text{ mm Hg}}\right)\left(\frac{1.0 \text{ cm H}_2\text{O}}{98.1 \text{ N/m}^2}\right) = \boxed{13.6 \text{ cm H}_2\text{O}}$$

(b) Solving this part in standard units, we know that: $P = P_0 + \Delta P = P_0 + \rho g h$, or

$$P = 1330 \text{ N/m}^2 + (1.05 \times 10^3 \text{ kg/m}^3)(9.80 \text{ m/s}^2)(0.600 \text{ m}) = 7504 \text{ N/m}^2,$$

then converting to cm water:

$$P = (7504 \text{ N/m}^2)\left(\frac{1.0 \text{ cm H}_2\text{O}}{98.1 \text{ N/m}^2}\right) = \boxed{76.5 \text{ cm H}_2\text{O}}$$

10.85 (a) Using Equation 10.3, we can calculate the pressure at a depth of 11.0 km: $P = h\rho g$, or

$$P = (11.0 \times 10^3 \text{ m})(1025 \text{ kg/m}^3)(9.80 \text{ m/s}^2) = 1.105 \times 10^8 \text{ N/m}^2 \times \frac{1 \text{ atm}}{1.013 \times 10^5 \text{ N/m}^2} = \boxed{1.09 \times 10^3 \text{ atm}}$$

(b) Using Equations 5.15 and 10.2:

$$\frac{\Delta V}{V_0} = \frac{1}{B}\frac{F}{A} = \frac{P}{B} = \frac{1.105 \times 10^8 \text{ N/m}^2}{2.2 \times 10^9 \text{ N/m}^2} = 5.02 \times 10^{-2} = \boxed{5.0\% \text{ decrease in volume}}.$$

(c) Using Equation 10.1, we can get an expression for percent change in density:

$$\frac{\Delta \rho}{\rho} = \frac{m/(V_0 - \Delta V)}{m/V_0} = \frac{V_0}{V_0 - \Delta V} = \frac{1}{1 - (\Delta V/V_0)} = \frac{1}{1.00 - 5.0 \times 10^{-2}} = 1.0526,$$

so that the percent increase in density is $\boxed{5.3\%}$. Therefore, the assumption of constant density is not strictly valid. The actual pressure would be greater, since the pressure is proportional to density.

11 FLUID DYNAMICS

CONCEPTUAL QUESTIONS

11.1 Flow rate, F , is the volume per unit time flowing past a point or through an area, *A*. Flow velocity is strictly how fast is a piece of the volume moving. Flow rate is proportional to the flow velocity, but it also depends on the cross-sectional area of the flow.

11.4 The broadening of the stream as it rises is due to slowing of the fluid. The equation of continuity shows us that as the velocity decreases, the area must increase, and so the stream broadens as it rises because its velocity slows due to gravity. Conversely, when the stream falls, its velocity increases due to gravity, and it narrows because of the equation of continuity. Surface tension should reduce the effect of the broadening while it increases the effect of the narrowing because the water wants to stay together.

11.7 A Wagner power painter is an example of entrainment. It pushes a high velocity air stream over the top of a tube of paint, sucking the paint in as droplets pushing it out the nozzle.

11.10 The height to which an entrainment device can raise a fluid is limited by the pressure difference between the high-velocity fluid and the entrained fluid.

11.13 The keel keeps the boat from tipping over when the wind blows in any direction other than straight ahead or straight behind.

11.16 The maximum height of the water squirting from leak 1 is the height of the water within the rubber boot. The speed of the water emerging from leak 2 is the same as that emerging from leak 1, only the direction is different (assuming the height of the leaks is the same). The velocity of the water is equal to the velocity the water would have, if it had dropped the distance from the top of the water within the boot to the leak. This can be seen by conservation of energy. The potential energy in the water at the top of the leak is converted into kinetic energy at the leak, and the maximum height of the water from leak 1 is then determined by converting that kinetic energy back into potential energy again.

11.19 Due to the viscosity of water, the water near the middle of a stream travels the fastest while the water near the shore travels the slowest. So, when traveling upstream, it is wisest to travel near the shore, where the current is slowest against you. Similarly, when traveling downstream, staying near the middle allows you to travel with the least amount of work because the current is fastest near the middle of the stream.

11.22 Blood velocity will be greatest where the pressure is the lowest, or at the constriction. The two distinct causes of higher resistance are the larger pressure drop in the constriction and the increase in turbulence.

11.25 Toward the back. When the car first slows, the air in the car compresses toward the front of the car, and for a split second the helium balloon will move forward. However, since the density of helium is smaller than that of air, the helium compresses the air in front of it less than the air next to the balloon compresses the air in front of and next to the balloon. Consequently the pressure directly in front of the balloon is slightly less than that to the slide, resulting to a backflow and the balloon goes backwards. Try it!

11.28 Osmosis is the transport of water through a semipermeable membrane from a region of high concentration to a region of low concentration. Dialysis is the transport of any other molecule through a semipermeable membrane due to its concentration difference. Both are transportation through semipermeable membranes due to concentration differences, but osmosis is for water while dialysis is for any other molecule.

PROBLEMS

Note: Throughout this chapter, Φ is the flow rate.

11.1 We are given the speed of the car and a gas mileage, giving us a volume consumed per time, so Equation 11.1,

$F = \dfrac{V}{t}$, is the formula we want to use to calculate the average flow rate:

$$\bar{F} = \frac{V}{t} = \frac{\text{speed}}{\text{gas mileage}} = \frac{60.0 \text{ mi/h}}{30.0 \text{ mi/gal}} \times \frac{3.785 \times 10^{-3} \text{ m}^3}{1 \text{ gal}} \times \frac{10^6 \text{ cm}^3}{1 \text{ m}^3} \times \frac{1 \text{ h}}{3600 \text{ s}} = \boxed{2.10 \text{ cm}^3/\text{s}}$$

11.7 (a) We are given the volume of the pool, and the flow rate, so using Equation 11.1, we can calculate the time to fill the pool:

$$F = \frac{V}{t} \Rightarrow t = \frac{V}{F} = \frac{20,000 \text{ gal}}{15.0 \text{ gal/min}} = 1.33 \times 10^3 \text{ min} = \boxed{22.2 \text{ h}}$$

(b) Using Equation 11.1 again, but with a new flow rate gives:

$$t = \frac{V}{F} = \frac{20,000 \text{ gal}}{50,000 \text{ ft}^3/\text{s}} \times \left(\frac{1 \text{ ft}}{0.3048 \text{ m}} \right)^3 \times \frac{3.785 \times 10^{-3} \text{ m}^3}{1 \text{ gal}} = \boxed{5.35 \times 10^{-2} \text{ s}} \text{ !}$$

11.13 If the fluid is incompressible, that tells us that the flow rate through both sides will be equal: $F = A_1 \bar{v}_1 = A_2 \bar{v}_2$, and writing the areas in terms of the diameter of the tube gives:

$$\pi \frac{d_1^2}{4} \bar{v}_1 = \pi \frac{d_2^2}{4} \bar{v}_2 \Rightarrow \bar{v}_2 = \bar{v}_1 \left(d_1^2 / d_2^2 \right) = \bar{v}_1 \left(d_1 / d_2 \right)^2 .$$

Therefore, the velocity through section 2, equals the velocity through section 1 time the square of the ratio of the diameters of section 1 and section 2.

11.19 Ignoring turbulence, we can use Bernoulli's equation:

$$P_1 + \frac{1}{2} \rho v_1^2 + \rho g h_1 = P_2 + \frac{1}{2} \rho v_2^2 + \rho g h_2 ,$$

where the heights are the same: $h_1 = h_2$ because we are concerned about above and below a thin roof. The velocity inside the house is zero, so $v_1 = 0.0 \text{ m/s}$, while the speed outside the house is $v_2 = 45.0 \text{ m/s}$. The difference in pressures, $P_1 - P_2$, can then be found: $P_1 - P_2 = \frac{1}{2} \rho v_2^2$. Now, we can relate the change in pressure to the force on the roof, using Equation 10.2, because we know the area of the roof ($A = 200 \text{ m}^2$):

$$F = (P_1 - P_2) A = \frac{1}{2} \rho \left(v_2^2 - v_1^2 \right) A$$

and substituting in the values gives:

$$F = \frac{1}{2} (1.14 \text{ kg/m}^3) \left[(45.0 \text{ m/s})^2 - (0.0 \text{ m/s})^2 \right] (220 \text{ m}^2) = \boxed{2.54 \times 10^5 \text{ N}} .$$

This extremely large force is the reason you should leave windows open in your house when there are tornadoes or heavy windstorms in the area, otherwise your roof will pop off!

11.25 Using Equation 11.6, we can calculate the power output by the left ventricle during the heartbeat:

$$\text{power} = \left(P + \frac{1}{2}\rho v^2 + \rho gh\right) \cdot F \text{ , where } P = 110 \text{ mm Hg} \times \frac{133 \text{ N/m}^2}{1.0 \text{ mm Hg}} = 1.463 \times 10^4 \text{ N/m}^2 \text{ ,}$$

$$\frac{1}{2}\rho v^2 = \frac{1}{2}\left(1.05 \times 10^3 \text{ kg/m}^3\right)\left(0.300 \text{ m/s}\right)^2 = 47.25 \text{ N/m}^2 \text{ , and}$$

$$\rho gh = \left(1.05 \times 10^3 \text{ kg/m}^3\right)\left(9.80 \text{ m/s}^2\right)\left(0.0500 \text{ m}\right) = 514.5 \text{ N/m}^2 \text{ , giving:}$$

$$\text{power} = \left(1.463 \times 10^4 \text{ N/m}^2 + 47.25 \text{ N/m}^2 + 514.5 \text{ N/m}^2\right)\left(83.0 \text{ cm}^3/\text{s}\right) \times \frac{10^{-6} \text{ m}^3}{\text{cm}^3} = \boxed{1.26 \text{ W}}$$

11.31 If the flow rate is reduced to 1.00% of its original value, then $F_2 = \dfrac{\Delta P \pi r_2^4}{8\eta l_2} = 0.0100 F_1 = 0.0100 \dfrac{\Delta P \pi r_1^4}{8\eta l_1}$. Since

the length of the arterioles is kept constant and the pressure difference is kept constant, we can get a relationship between the radii:

$$r_2^4 = 0.0100 r_1^4 \Rightarrow r_2 = \left(0.0100\right)^{1/4} r_1 = \boxed{0.316 r_1} \text{ .}$$

The radius is reduced to 31.6% of the original radius to reduce the flow rate to 1.00% of its original value.

11.37 (a) We can calculate the pressure using Equation 10.2, where the height is 1.61 m and the density is that of seawater:

$$P_2 = \rho hg = \left(1025 \text{ kg/m}^3\right)\left(1.61 \text{ m}\right)\left(9.80 \text{ m/s}^2\right) = \boxed{1.62 \times 10^4 \text{ N/m}^2}$$

(b) If the pressure is decreased to 1.50 m, we use Equation 11.10 to determine the new flow rate:

$$F = \frac{\left(P_2 - P_1\right)\pi r^4}{8\eta l} \text{ . We use } l = 0.0250 \text{ m, } r = 0.150 \times 10^{-3} \text{ m, } \eta = 1.005 \times 10^{-3} \text{ N} \cdot \text{s/m}^2 \text{ , and}$$

$$P_1 = 1.066 \times 10^3 \text{ N/m}^2 \text{ . Using Equation 10.2, we can find the pressure due to a depth of 1.50 m:}$$

$$P_2' = \left(1025 \text{ kg/m}^3\right)\left(1.50 \text{ m}\right)\left(9.80 \text{ m/s}^2\right) = 1.507 \times 10^4 \text{ N/m}^2 \text{ . So substituting into Equation 11.10 gives:}$$

$$F = \frac{\left(1.507 \times 10^4 \text{ N/m}^2 - 1.066 \times 10^3 \text{ N/m}^2\right)\pi\left(0.150 \times 10^{-3} \text{ m}\right)^4}{8\left(1.005 \times 10^{-3} \text{ N} \cdot \text{s/m}^2\right)\left(0.0250 \text{ m}\right)} = 1.11 \times 10^{-7} \text{ m}^3/\text{s} = \boxed{0.111 \text{ cm}^3/\text{s}}$$

(c) The flow rate will be zero (and become negative) when the pressure in the IV is equal to (or less than) the pressure in the patient's vein: $P_r = \rho hg \Rightarrow h = \dfrac{P_r}{\rho g} = \dfrac{1.066 \times 10^3 \text{ N/m}^2}{\left(1025 \text{ kg/m}^3\right)\left(9.80 \text{ m/s}^2\right)} = 0.106 \text{ m} = \boxed{10.6 \text{ cm}}$.

11.43 Since we will use Equation 11.12 to determine the Reynold's number: $N_R = \dfrac{2\rho vr}{\eta}$, we first must determine the

velocity of the oil. Since the oil rises to 25.0 m, use kinematics, $v^2 = v_0^2 - 2gy$, where $v = 0$ m/s , $y = 25.0$ m ,

and find v_0 : $v_0 = \sqrt{2gy} = \sqrt{2\left(9.80 \text{ m/s}^2\right)\left(25.0 \text{ m}\right)} = 22.136 \text{ m/s}$. Now, we can use Equation 11.12:

$$N_R = \frac{2\left(900 \text{ kg/m}^3\right)\left(22.136 \text{ m/s}\right)\left(0.0500 \text{ m}\right)}{1.00 \left(\text{N/m}^2\right) \cdot \text{s}} = \boxed{1.99 \times 10^3} < 2000 \text{ ,}$$

where $N_R = 2000$ is the approximate upper value for laminar flow. So the flow of oil is laminar (barely).

11.49 (a) We will use Equation 11.12, where $N_R = \dfrac{2\rho v r}{\eta} \leq 1000$, to find the minimum radius, which will give us the

minimum diameter. First, we need to get an expression for the velocity, from Equation 11.2: $v = \dfrac{F}{A} = \dfrac{F}{\pi r^2}$.

Substituting into Equation 11.12 gives:

$$\frac{2\rho\left(F/\pi r^2\right)r}{\eta} = \frac{2\rho F}{\pi \eta r} \leq 1000 \text{ or } r \geq \frac{2\rho F}{\pi \eta (1000)},$$

so that the minimum diameter is:

$$d \geq \frac{4\rho F}{1000 \pi \eta} = \frac{4\left(680 \text{ kg/m}^3\right)\left(3.00 \times 10^{-2} \text{ m}^3/\text{s}\right)}{1000\pi\left(1.00 \times 10^{-3} \text{ N} \cdot \text{s/m}^2\right)} = 2 \times 12.99 \text{ m} = \boxed{26.0 \text{ m}}$$

(b) Using Equation 11.10, $F = \dfrac{\Delta P \pi r^4}{8\eta l}$, we can determine the pressure difference from the flow rate:

$$\Delta P = \frac{8\eta l F}{\pi r^4} = \frac{8\left(1.00 \times 10^{-3} \text{ N} \cdot \text{s/m}^2\right)(1000 \text{ m})\left(3.00 \times 10^{-2} \text{ m}^3/\text{s}\right)}{\pi (12.99 \text{ m})^4} = \boxed{2.68 \times 10^{-6} \text{ Pa}}.$$

This pressure is equivalent to 2.65×10^{-11} atm, which is a very small pressure!

11.55 From Table 11.2, we know: $D_{H_2} = 6.4 \times 10^{-5} \text{ m}^2/\text{s}$, and $D_{O_2} = 1.8 \times 10^{-5} \text{ m}^2/\text{s}$. We want to use Equation 11.14, since that relates the time to the distance traveled during diffusion. We have two equations: $x_{\text{rms}, O_2} = \sqrt{2D_{O_2} t_{O_2}}$, and $x_{\text{rms}, H_2} = \sqrt{2D_{H_2} t_{H_2}}$. We want the distance traveled to be the same, so we can set the equations equal. The distance will be the same when the time difference between t_{H_2} and t_{O_2} is 1.00 s, so we can relate the two times: $t_{O_2} = t_{H_2} + 1.00 \text{ s}$. Setting the two distance equations equal and squaring gives: $2D_{O_2} t_{O_2} = 2D_{H_2} t_{H_2}$, and substituting for the oxygen time gives: $D_{O_2}\left(t_{H_2} + 1.00 \text{ s}\right) = D_{H_2} t_{H_2}$. Solving for the hydrogen time gives:

$$t_{H_2} = \frac{D_{O_2}}{D_{H_2} - D_{O_2}} \times 1.00 \text{ s} = \frac{1.8 \times 10^{-5} \text{ m}^2/\text{s}}{6.4 \times 10^{-5} \text{ m}^2/\text{s} - 1.8 \times 10^{-5} \text{ m}^2/\text{s}} \times 1.00 \text{ s} = \boxed{0.391 \text{ s}}.$$

The hydrogen will take 0.391 s to travel to the distance x, while the oxygen will take 1.391 s to travel the same distance. Therefore, the hydrogen will be 1.00 seconds ahead of the oxygen after 0.391 s.

12 TEMPERATURE, KINETIC THEORY, AND THE GAS LAWS

CONCEPTUAL QUESTIONS

12.1 Systems are in thermal equilibrium when they have the same temperature.

12.4 Since the volume is kept constant, when the temperature changes the pressure changes, so the pressure of the gas is measured to indicate its temperature.

12.7 There will be significant cell damage, on the order of 10% to 30%, if the body is frozen, so the prospect of freezing and then thawing a person at some future date is not very good. The body must be able to withstand the loss of approximately 10-30% of its cells and still be able to survive.

12.10 If the pressure is kept at atmospheric pressure, then the volume of a gas would be zero at 0 K, based on the ideal gas law. This will not happen because the gas would have changed phase before it reached 0 K, and the ideal gas law would no longer apply. Physically the volume could not be zero because the molecules themselves in the gas take up physical space and cannot be compressed to zero volume.

12.13 At higher pressures, the boiling point of water is at a higher temperature. In a pressure cooker, the sustainable temperature is higher than without the increased pressure, leading to an increased cooking speed.

12.16 Yes, carbon dioxide can be liquefied at room temperature (20°C), if the pressure is raised to 56 atm. This is a rather high pressure, but it can be obtained.

12.19 Yes, the saturation vapor densities listed in Table 12.4 will be valid in an atmosphere of helium because the vapor density is independent of the presence of other gases. At high altitudes, the condensation rate decreases because of decreased pressure, and therefore the vapor density is larger.

PROBLEMS

12.1 Using Equation 12.1a, we can convert from Celsius to Fahrenheit:

$$T(°F) = \frac{9}{5}T(°C) + 32.0° = \frac{9}{5}(39.0°) + 32.0° = \boxed{102°F}.$$

So 39.0°C is equivalent to 102°F .

12.7   (a)   We can use Equation 12.1b, $T\left(^{\circ}C\right)=\dfrac{5}{9}\left(T\left(^{\circ}F\right)-32^{\circ}\right)$, to determine the change in temperature.

$$\Delta T\left(^{\circ}C\right)=T_2\left(^{\circ}C\right)-T_1\left(^{\circ}C\right)=\frac{5}{9}\left[T_2\left(^{\circ}F\right)-T_1\left(^{\circ}F\right)\right]=\frac{5}{9}\Delta T\left(^{\circ}F\right)=\frac{5}{9}\left(40^{\circ}\right)=\boxed{22.2^{\circ}C}$$

(b)   Using Equation 12.1a: $T\left(^{\circ}F\right)=\dfrac{9}{5}T\left(^{\circ}C\right)+32.0^{\circ}$, we can determine the change in temperature:

$$\Delta T\left(^{\circ}F\right)=T_2\left(^{\circ}F\right)-T_1\left(^{\circ}F\right)=\frac{9}{5}\left[T_2\left(^{\circ}C\right)+32.0^{\circ}\right]-\frac{9}{5}\left[T_1\left(^{\circ}C\right)+32.0^{\circ}\right]=\frac{9}{5}\left[T_2\left(^{\circ}C\right)-T_1\left(^{\circ}C\right)\right]=\frac{9}{5}\Delta T\left(^{\circ}C\right).$$

We could have also used the equation we derived in part (a) and solved for $\Delta T\left(^{\circ}F\right)$, to get the same result!

12.13   Using Equation 12.4, we can get an expression for the change in volume: $\Delta V=\beta V_0\Delta T$, so the final volume is:
$V=V_0+\Delta V=V_0\left(1+\beta\Delta T\right)$, where $\beta=950\times10^{-6}$ /°C for gasoline (see Table 12.1), so that

$$V=\left(60.00\text{ L}\right)\left[1+\left(950\times10^{-6}/^{\circ}C\right)\left(35.0^{\circ}C-15.0^{\circ}C\right)\right]=\boxed{61.14\text{ L}}.$$

As the temperature is increased, the volume also increases.

12.19   Using Equation 12.2: we know how the length changes with temperature: $\Delta L=\alpha L_0\Delta T$. Also, we know that the volume of a cube is related to its length by: $V=L^3$, so the final volume is then: $V=V_0+\Delta V=\left(L_0+\Delta L\right)^3$. Substituting for ΔL gives:

$$V=V_0+\Delta V=\left(L_0+\alpha L_0\Delta T\right)^3=L_0^3\left(1+\alpha\Delta T\right)^3.$$

Now, because $\alpha\Delta T$ is small, we can use the binomial expansion:

$$V=V_0+\Delta V=L_0^3\left(1+\alpha\Delta T\right)^3\approx L_0^3\left(1+3\alpha\Delta T\right)=L_0^3+3\alpha L_0^3\Delta T,$$

so writing the length terms in terms of volumes gives: $V=V_0+\Delta V\approx V_0+3\alpha V_0\Delta T$, and so

$$\Delta V=\beta V_0\Delta T\approx3\alpha V_0\Delta T,\text{ or }\boxed{\beta\approx3\alpha}.$$

12.25   (a)   This is a units conversion problem, so

$$\frac{N}{V}=\left(2.68\times10^{25}/\text{m}^3\right)\left(\frac{1\text{ m}}{100\text{ cm}}\right)^3=\boxed{2.68\times10^{19}/\text{cm}^3}$$

(b)   Again, we need to convert the units:

$$\frac{N}{V}=\left(2.68\times10^{25}/\text{m}^3\right)\left(\frac{1\text{ m}}{1.00\times10^6\ \mu\text{m}}\right)^3=\boxed{2.68\times10^7\ /\mu\text{m}^3}$$

(c)   This says that atoms and molecules must be on the order of (if they were tightly packed):

$$V=\frac{N}{2.68\times10^7\ /\mu\text{m}^3}=\frac{1}{2.68\times10^7\ /\mu\text{m}^3}=3.73\times10^{-8}\ \mu\text{m}^3,$$

or the average length of an atom is less than approximately: $\left(3.73\times10^{-8}\ \mu\text{m}^3\right)^{1/3}=3.34\times10^{-3}\ \mu\text{m}$ $\boxed{3\text{ nm}}$.
Since atoms are widely spaced, the average length is probably more on the order of 0.3 nm.

12.31 First, we need to convert the temperature, and volume to standard units:

$$T(K) = T(°C) + 273.15 = 18.0 + 273.15 = 291.2 \text{ K}, \text{ and } V = 2.00 \text{ L} \times \frac{10^{-3} \text{ m}^3}{L} = 2.00 \times 10^{-3} \text{ m}^3.$$

Next, use the ideal gas law, Equation 12.5: to determine the initial number of molecules in the tire:

$$P_1 V = N_1 kT \Rightarrow N_1 = \frac{P_1 V}{kT} = \frac{(7.00 \times 10^5 \text{ N/m}^2)(2.00 \times 10^{-3} \text{ m}^3)}{(1.38 \times 10^{-23} \text{ J/K})(291.15 \text{ K})} = 3.484 \times 10^{23}.$$

Then, we need to determine how many molecules were removed from the tire. Use the ideal gas law, with the volume released and atmospheric pressure:

$$PV = \Delta N kT \Rightarrow \Delta N = \frac{PV}{kT} = \frac{(1.013 \times 10^5 \text{ N/m}^2)\left(100 \text{ cm}^3 \times \frac{10^{-6} \text{ m}^3}{\text{cm}^3}\right)}{(1.38 \times 10^{-23} \text{ J/K})(291.15 \text{ K})} = 2.521 \times 10^{21}.$$

So, we can now determine how many molecules remain after the gas is released:

$$N_2 = N_1 - \Delta N = 3.484 \times 10^{23} - 2.521 \times 10^{21} = 3.459 \times 10^{23}.$$

Finally, we can determine the final pressure:

$$P_2 = \frac{N_2 kT}{V} = \frac{(3.459 \times 10^{23})(1.38 \times 10^{-23} \text{ J/K})(291.15 \text{ K})}{2.00 \times 10^{-3} \text{ m}^3} = \boxed{6.95 \times 10^5 \text{ N/m}^2}$$

12.37 (a) Use the ideal gas law, where $\rho = \dfrac{N}{V} = 10^6 \text{ /m}^3$,

$$PV = NkT \Rightarrow P = \frac{N}{V} kT = \frac{10^6}{1 \text{ m}^3}(1.38 \times 10^{-23} \text{ J/K})(2.7 \text{ K}) = 3.73 \times 10^{-17} \text{ N/m}^2 = \boxed{3.7 \times 10^{-17} \text{ N/m}^2}$$

(b) Now, using the pressure found in part (a), let $n = 1.00 \text{ mol}$, and use the ideal gas law, Equation 12.8:

$$PV = nRT \Rightarrow V = \frac{nRT}{P} = \frac{(1.00 \text{ mol})(8.31 \text{ J/mol} \cdot \text{K})(2.7 \text{ K})}{3.73 \times 10^{-17} \text{ N/m}^2} = 6.02 \times 10^{17} \text{ m}^3 = \boxed{6.0 \times 10^{17} \text{ m}^3}$$

(c) Since the volume of a cube is its length cubed: $L = V^{1/3} = (6.02 \times 10^{17} \text{ m}^3)^{1/3} = 8.45 \times 10^5 \text{ m} = \boxed{8.4 \times 10^2 \text{ km}}$

12.43 Use Equation 12.10 to find the temperature:

$$\overline{KE} = \frac{3}{2}kT \Rightarrow T = \frac{2\overline{KE}}{3k} = \frac{2(6.40 \times 10^{-14} \text{ J})}{3(1.38 \times 10^{-23} \text{ J/K})} = \boxed{3.09 \times 10^9 \text{ K}}$$

12.49 (a) Using Table 12.4: vapor pressure$_{H_2O}$ (20°C) = $\boxed{2.33 \times 10^3 \text{ N/m}^2}$

(b) Divide the vapor pressure by atmospheric pressure: $\dfrac{2.33 \times 10^3 \text{ N/m}^2}{1.013 \times 10^5 \text{ N/m}^2} \times 100\% = \boxed{2.30\%}$

(c) Use the saturation vapor density for water from Table 12.4, and divide by the density of dry air to get the percent of water in the air: $\dfrac{17.2 \times 10^{-3} \text{ kg/m}^3}{1.20 \text{ kg/m}^3} \times 100\% = \boxed{1.43\%}$

12.55 Using Equation 12.12, $\text{percent relative humidity} = \dfrac{\text{vapor density}}{\text{saturation vapor density}} \times 100\%$, where

percent relative humidity $= 6.00\%$, and saturation vapor density $= 51.1 \ \text{g/m}^3$ (for water at $40°C$, from Table 12.4), giving:

$$\text{vapor density} = \frac{(\text{percent relative humidity})(\text{saturation vapor density})}{100\%} = \frac{(6.00\%)(51.1 \ \text{g/m}^3)}{100\%} = \boxed{3.07 \ \text{g/m}^3}$$

12.61 (a) The partial pressure is the pressure a gas would create if its alone occupied the total volume, or the partial pressure is the percent the gas occupies times the total pressure:

$$\text{partial pressure}(O_2) = (\% \ O_2)(\text{atmospheric pressure}) = (0.209)(3.30 \times 10^4 \ \text{N/m}^2) = \boxed{6.90 \times 10^3 \ \text{N/m}^2}$$

(b) First calculate the partial pressure at sea level:

$$\text{partial pressure}(@ \text{ sea level}) = (0.209)(1.013 \times 10^5 \ \text{N/m}^2) = 2.117 \times 10^4 \ \text{N/m}^2 ,$$

then set that equal to the percent oxygen times the pressure at the top of Mt. Everest:

$$\text{partial pressure } (@ \text{ sea level}) = \frac{\% \ O_2}{100\%}(3.30 \times 10^4 \ \text{N/m}^2) = 2.117 \times 10^4 \ \text{N/m}^2 ,$$

so that

$$\% \ O_2 = \frac{2.117 \times 10^4 \ \text{N/m}^2}{3.30 \times 10^4 \ \text{N/m}^2} \times 100\% = \boxed{64.2\%} .$$

The mountain climber should breath air containing 64.2% oxygen at the top of Mt. Everest to maintain the same partial pressure as at sea level. Clearly, the air does not contain that much oxygen. This is why you feel lightheaded at high altitudes; you are partially oxygen deprived!

(c) This drying process occurs because the partial pressure of water molecules at high altitudes is decreased substantially, and the breathing passages are therefore not getting the moisture they require from the air being breathed.

12.67 (a) From Table 12.4, we can get the vapor pressure of water at $150°C$: vapor pressure $= \boxed{4.76 \times 10^5 \ \text{N/m}^2}$.

(b) Using Equation 10.2, $P = \dfrac{F}{A}$, we can calculate the force exerted on the pressure cooker lid:

$$F = P \cdot A = P \cdot \pi r^2 = 4.76 \times 10^5 \ \text{N/m}^2 \cdot \pi (0.125 \ \text{m})^2 = \boxed{2.34 \times 10^4 \ \text{N}}$$

13 HEAT AND HEAT TRANSFER METHODS

CONCEPTUAL QUESTIONS

13.1 Heat is energy transferred solely because of a temperature difference.

13.4 The three factors that affect the amount of heat needed to change an object's temperature are its specific heat in all phases (gas, liquid and solid), its latent heat of fusion, and its latent heat of vaporization.

13.7 Phase changes can have a tremendous stabilizing effect even on temperatures not near the melting and boiling points. The latent heat of fusion of water, being 334 kJ/kg, means that water gives off a large amount of heat when it turns from water to ice. This heat helps slow the decrease of air temperature near large bodies of water, preventing them from falling significantly below 0°C.

13.10 Condensation forms on a glass of ice water because the temperature of nearby air is reduced below the dew point. The air cannot hold as much water as it did at room temperature, so water condenses. Heat of vaporization is released when the water condenses, speeding the melting of the ice in the glass.

13.13 0°C ice requires a large amount of heat to transform it to 0°C water, before its temperature can be raised above 0°C. So, this additional heat that can be absorbed by the ice is the reason that 0°C ice stays cold much longer than 0°C water in an ice bag.

13.16 When the temperature is at the dew point, water vapor from the air condenses, giving off heat. This extra heat keeps the temperature of the air from dropping below the dew point.

13.19 Conduction is the major method of heat transfer from the hot core of the earth to its surface. Radiation is the major method of heat transfer from the surface of the earth to outer space.

13.22 The external air supply brings in more oxygen allowing the coals to burn hotter. This in turn allows radiation to transfer more heat to the room, making the fire more energy efficient. The fire heats the circulating room air by conduction, and it carries that heat out to the room by convection.

13.25 The large trees reflect the radiated heat from the horses' bodies back to the horses, keeping them warmer.

13.28 At night, clouds reflect the radiated heat of the earth back to the surface, greatly reducing heat loss, just as they greatly reduce heat gain during the day, keeping it warmer than if there were no clouds.

PROBLEMS

13.1   The heat input is given by Equation 13.2, where the specific heat of water is $c = 4186$ J/kg$\cdot$°C, the mass is given by Equation 10.1:

$$m = \rho V = \left(1.00 \times 10^3 \text{ kg/m}^3\right)\left(80,000 \text{ L}\right) \times \frac{1 \text{ m}^3}{1000 \text{ L}} = 8.00 \times 10^4 \text{ kg},$$

and the temperature change is $\Delta T = 1.50$°C. Therefore,

$$Q = mc\Delta T = \left(8.00 \times 10^4 \text{ kg}\right)\left(4186 \text{ J/kg}\cdot\text{°C}\right)\left(1.50\text{°C}\right) = \boxed{5.02 \times 10^8 \text{ J}}$$

13.7   (a)   Use Equation 13.2, since there is no phase change involved in heating the water:

$$Q = mc\Delta T = \left(0.800 \text{ kg}\right)\left(4186 \text{ J/kg}\cdot\text{°C}\right)\left(30.0\text{°C}\right) = \boxed{1.00 \times 10^5 \text{ J}}$$

(b)   To determine the heat required, we must melt the ice, using Equation 13.3, and then add the heat required to raise the temperature of the melted ice using Equation 13.2, so that

$$Q = mL_f + mc\Delta T = \left(0.800 \text{ kg}\right)\left(334 \times 10^3 \text{ J/kg}\right) + 1.005 \times 10^5 \text{ J} = \boxed{3.68 \times 10^5 \text{ J}}$$

(c)   The ice is much more effective in absorbing heat because it first must be melted, which requires quite a lot of energy, then it gains the heat that the water also would. The first 2.67×10^5 J of heat is used to melt the ice, then it absorbs the 1.00×10^5 J of heat that the water absorbs.

13.13   (a)   The heat required can be calculated in three steps: first, heat the water to 100°C (using Equation 13.2), then turn the water into steam at 100°C (using Equation 13.3), and finally heat the steam to 450°C (using Equation 13.2, being careful to use the specific heat of steam here):

$$Q = mc_w\Delta T_{30-100} + mL_v + mc_s\Delta T_{100-450}$$

So that:

$$Q = \left(1000 \text{ kg}\right)\left(4186 \text{ J/kg}\cdot\text{°C}\right)\left(100\text{°C} - 30.0\text{°C}\right) + \left(1000 \text{ kg}\right)\left(2256 \times 10^3 \text{ J/kg}\right)$$

$$+ \left(1000 \text{ kg}\right)\left(1520 \text{ J/kg}\cdot\text{°C}\right)\left(450\text{°C} - 100\text{°C}\right) = \boxed{3.08 \times 10^9 \text{ J}}$$

(b)   We calculate the power using Equation 6.9, where the heat provides the work done:

$$P = \frac{Q}{t} = \frac{3.08 \times 10^9 \text{ J}}{1.00 \text{ s}} = \boxed{3.08 \times 10^3 \text{ MW}}$$

13.19   The heat gained in evaporating the coffee equals the heat leaving the coffee and glass to lower its temperature, so that:

$$ML_v = m_c c_c \Delta T + m_g c_g \Delta T,$$

where $M =$ mass of coffee that evaporates. Solving for the evaporated coffee gives:

$$M = \frac{\Delta T \left(m_c c_c + m_g c_g\right)}{L_v} = \frac{\left(95.0\text{°C} - 45.0\text{°C}\right)\left[\left(350 \text{ g}\right)\left(1.00 \text{ cal/g}\cdot\text{°C}\right) + \left(100 \text{ g}\right)\left(0.20 \text{ cal/g}\cdot\text{°C}\right)\right]}{560 \text{ cal/g}} = \boxed{33.0 \text{ g}}.$$

Notice that we did the problem in calories and grams, since the latent heat was given in those units, and the result we wanted was in grams. We could have done the problem in standard units, and then converted back to grams to get the same answer.

13.25 The heat gained by the ice equals the heat lost by the water. Since we do not know the final state of the water/ice combination, we first need to compare the heat needed to raise the ice to 0°C and the heat available from the water: First, we need to calculate how much heat would be required to raise the temperature of the ice to 0°C:

$$Q_{ice} = mc\Delta T = (1.20 \text{ kg})(2090 \text{ J/kg} \cdot °C)(15°C) = 3.762 \times 10^4 \text{ J}.$$

Now, we need to calculate how much heat is given off to lower the water to 0°C:

$$Q_1 = mc\Delta T_1 = (0.0100 \text{ kg})(4186 \text{ J/kg} \cdot °C)(20.0°C) = 837.2 \text{ J},$$

since this is less than the heat required to heat the ice, we need to calculate how much heat is given off to convert the water to ice:

$$Q_2 = mL_f = (0.0100 \text{ kg})(334 \times 10^3 \text{ J/kg}) = 3.340 \times 10^3 \text{ J},$$

so the total amount of heat given off to turn the water to ice at 0°C is: $Q_{water} = 4.177 \times 10^3 \text{ J}$. Since $Q_{ice} > Q_{water}$, we have determined that the final state of the water/ice is ice, at some temperature below 0°C.

Now, we need to calculate the final temperature. We use the heat lost from the water equals the heat gained by the ice, where we now know that the final state is ice at $T_f < 0°C$: $Q_{lost\ by\ water} = Q_{gained\ by\ ice}$, or

$$m_{water}c_{water}\Delta T_{20 \to 0} + m_{water}L_f + m_{water}c_{ice}\Delta T_{0 \to ?} = m_{ice}c_{ice}\Delta T_{-15 \to ?}.$$

Substituting for the change in temperatures (being careful that ΔT is always positive) and simplifying gives:

$$m_{water}\left[(c_{water})(20°C) + L_f + (c_{ice})(0 - T_f)\right] = m_{ice}c_{ice}\left[T_f - (-15°C)\right].$$

Solving for the final temperature gives: $T_f = \dfrac{m_{water}\left[(c_{water})(20°C) + L_f\right] - m_{ice}c_{ice}(15°C)}{(m_{water} + m_{ice})c_{ice}}$, and so finally:

$$T_f = \frac{(0.0100 \text{ kg})\left[(4186 \text{ J/kg} \cdot °C)(20°C) + 334 \times 10^3 \text{ J}\right] - (1.20 \text{ kg})(2090 \text{ J/kg} \cdot °C)(15°C)}{(0.0100 \text{ kg} + 1.20 \text{ kg})(2090 \text{ J/kg} \cdot °C)} = \boxed{-13.2°C}.$$

13.31 First, calculate how much heat must be dissipated:

$$Q = mc_{human\ body}\Delta T = (80.0 \text{ kg})(3500 \text{ J/kg} \cdot °C)(40.0°C - 37.0°C) = 8.40 \times 10^5 \text{ J}.$$

Then, since power is heat divided by time, we can get the power required to produce the calculated amount of heat in 30.0 minutes:

$$P_{cooling} = \frac{Q}{t} = \frac{8.40 \times 10^5 \text{ J}}{(30 \text{ min})(60 \text{ s/1 min})} = 4.67 \times 10^2 \text{ W}.$$

Now, since the body continues to produce heat at a rate of 150 W, we need to add that to the required cooling power:

$$P_{required} = P_{cooling} + P_{body} = 467 \text{ W} + 150 \text{ W} = \boxed{617 \text{ W}}.$$

13.37 Use Equation 13.3, where $Q = \dfrac{1}{2} 3000 \text{ kcal}$ and $L_{v(37°C)} = 580 \text{ kcal/kg}$, from the footnote on Table 13.2:

$$Q = mL_{v(37°C)} \Rightarrow m = \frac{Q}{L_v} = \frac{1500 \text{ kcal}}{580 \text{ kcal/kg}} = \boxed{2.59 \text{ kg}}$$

13.43 (a) Using the Stefan-Boltzmann law of radiation (Equation 13.6), making sure to convert the temperatures into units of Kelvin, the rate of heat transfer is:

$$\frac{Q}{t} = \sigma e A\left(T_2^4 - T_1^4\right) = \left(5.67 \times 10^{-8} \ \text{J/s} \cdot \text{m}^2 \cdot \text{K}^4\right)(0.750)(1.20 \ \text{m}^2)\left[(323 \ \text{K})^4 - (383 \ \text{K})^4\right] = \boxed{-543 \ \text{W}}$$

(b) Assuming an automobile engine is 100 horsepower and the efficiency of a gasoline engine is 30% (from Table 6.3), the engine consumes $\dfrac{100 \ \text{horsepower}}{30\%} = 333.3$ horsepower in order to generate the 100 horsepower. Therefore, 233.3 horsepower is lost due to heating. Since 1 hp = 746 W, the radiator transfers $543 \ \text{W} \times \dfrac{1 \ \text{hp}}{746 \ \text{W}} = 0.728$ hp from radiation, which is not a significant fraction because the heat is primarily transferred from the radiator by other means.

13.49 Use the rate of heat transfer by conduction, Equation 13.4, $\dfrac{Q}{t} = \dfrac{kA(T_2 - T_1)}{d}$, and take the ratio of the wall to the window. The temperature difference for the wall and window will be the same:

$$\frac{(Q/t)_{\text{wall}}}{(Q/t)_{\text{window}}} = \frac{k_{\text{wall}} A_{\text{wall}} d_{\text{window}}}{k_{\text{window}} A_{\text{window}} d_{\text{wall}}} = \frac{(2 \times 0.042 \ \text{J/s} \cdot \text{m} \cdot {}^\circ\text{C})(10.0 \ \text{m}^2)(0.750 \times 10^{-2} \ \text{m})}{(0.84 \ \text{J/s} \cdot \text{m} \cdot {}^\circ\text{C})(2.00 \ \text{m}^2)(13.0 \times 10^{-2} \ \text{m})} = \boxed{0.029 \ \text{wall-to-window}},$$

so windows conduct more heat than walls. This should seem reasonable, since in the winter the windows feel colder than the walls.

13.55 Use Equation 13.2, in combination with Equations 6.9 and 10.1 to get: $P = \dfrac{Q}{t} = \dfrac{mc\Delta T}{t} = \dfrac{\rho V c \Delta T}{t}$. Now, since $c = 721 \ \text{J/kg} \cdot {}^\circ\text{C}$ (from Table 13.1), and $\rho = 1.29 \ \text{kg/m}^3$ (from Table 10.1), we get:

$$P = \frac{Q}{t} = \frac{\rho V c \Delta T}{t} = \frac{\left(1.29 \ \text{kg/m}^3\right)\left(500 \ \text{m}^3\right)\left(721 \ \text{J/kg} \cdot {}^\circ\text{C}\right)(10.0^\circ\text{C})}{60.0 \ \text{s}} = 7.75 \times 10^4 \ \text{W} = \boxed{77.5 \ \text{kW}}.$$

13.61 *Step 1*: Use the rate of heat transfer.
Step 2: The type of heat transfer is conduction.

Step 3: Find: (a) $\dfrac{Q}{t}$, (b) t.

Step 4: Given: (a) $k = 0.84 \ \text{J/s} \cdot \text{m} \cdot {}^\circ\text{C}$ (from Table 13.3), $A = 1.00 \ \text{m}^2$, $\Delta T = 30.0^\circ\text{C}$, $d = 30.0 \ \text{m}$;
    (b) $m = 2700 \ \text{kg}$, $c = 840 \ \text{J/kg} \cdot {}^\circ\text{C}$ (from Table 13.1), and $\Delta T = 1.00^\circ\text{C}$.

Step 5: (a) Use Equation 13.4: $P = \dfrac{Q}{t} = \dfrac{kA\Delta T}{d}$, (b) Use Equation 13.2 $Q = Pt = mc\Delta T \Rightarrow t = \dfrac{mc\Delta T}{P}$.

Step 6: Substitute in the values:

(a) $P = \dfrac{Q}{t} = \dfrac{kA\Delta T}{d} = \dfrac{(0.84 \ \text{J/s} \cdot \text{m} \cdot {}^\circ\text{C})(1.00 \ \text{m}^2)(30.0^\circ\text{C})}{30.0 \ \text{m}} = \boxed{0.84 \ \text{W}}$

(b) $t = \dfrac{mc\Delta T}{P} = \dfrac{(2700 \ \text{kg})(840 \ \text{J/kg} \cdot {}^\circ\text{C})(1.00^\circ\text{C})}{0.84 \ \text{W}} = 2.7 \times 10^6 \ \text{s} \times \dfrac{1 \ \text{d}}{8.64 \times 10^4 \ \text{s}} = \boxed{31 \ \text{d}}$

Step 7: These answers seem reasonable, given the fact that lava beds and slag heaps cool over the period of years.

13.67 (a) Use Equation 13.6: $\dfrac{Q}{t} = \sigma e\left(\dfrac{A}{2}\right)\left(T_2^4 - T_1^4\right)$, gives: $T_2 = \left[\dfrac{Q/t}{\sigma e(A/2)} + T_1^4\right]^{1/4} = \left[\dfrac{2(Q/t)}{\sigma e A} + T_1^4\right]^{1/4}$, so that:

$$T_2 = \left[\frac{2(20.0\text{ W})}{\left(5.67\times10^{-8}\text{ J/s}\cdot\text{m}^2\cdot\text{K}^4\right)(0.970)\left(0.400\text{ m}^2\right)} + (307\text{ K})^4\right]^{1/4} = 321.63\text{ K} = \boxed{48.5°\text{C}}$$

(b) Pure white objects reflect more of the radiant energy that hits it, so the white tent would prevent more of the sunlight from heating up the inside of the tent, and the white tunic would prevent that which entered the tent from heating the rider. Therefore, with a white tent, the temperature would be lower than $48.5°\text{C}$, and the rate of radiant heat transferred to the rider would be smaller than 20.0 W.

13.73 (a) From Table 6.4, you produce power at a rate of 685 W, and since you are 20% efficient, you must have

generated: $P_{\text{generated}} = \dfrac{P_{\text{produced}}}{\text{efficiency}} = \dfrac{685\text{ W}}{0.20} = 3425\text{ W}$. If only 685 W of power was useful, the power

available to heat the body is $P_{\text{wasted}} = 3425\text{ W} - 685\text{ W} = 2.74\times10^3\text{ W}$. Now, using Equation 13.2,

$Q = mc\Delta T$, we can determine how long it takes to create $\Delta T = 1.00°\text{C}$, because $P_{\text{wasted}} = \dfrac{Q}{t} = \dfrac{mc\Delta T}{t}$. We

know $c = 3500\text{ J/kg}\cdot°\text{C}$ (from Table 13.1), so that

$$t = \frac{mc\Delta T}{P_{\text{wasted}}} = \frac{(76.0\text{ kg})(3500\text{ J/kg}\cdot°\text{C})(1.00°\text{C})}{2.74\times10^3\text{ W}} = \boxed{97.1\text{ s}}$$

(b) This says that it takes about a minute and a half to generate enough heat to raise the temperature of your body by $1.00°\text{C}$, which seems quite reasonable. Generally, within five minutes of working out on a stairmaster, you definitely feel warm and probably are sweating to keep your body from overheating!

13.79 (a) Use Equation 13.2, $Q = mc\Delta T$, where $Q = (0.950)(\text{food consumed})$, so that

$$\Delta T = \frac{Q}{mc} = \frac{(0.950)(2500\text{ kcal})}{(80.0\text{ kg})(0.83\text{ kcal}\cdot°\text{C})} = \boxed{36°\text{C}}.$$

This says that the temperature of the person is $37°\text{C} + 36°\text{C} = 73°\text{C}$!

(b) This temperature is way too high! The temperature of a human being should remain below $40°\text{C}$.

(c) The assumption that the person retains 95% of the energy as body heat is unreasonable. Most of the food consumed in a day is converted to body heat, losing energy by sweating and breathing, etc.

THERMODYNAMICS

14

CONCEPTUAL QUESTIONS

14.1 The first law of thermodynamics is conservation of internal energy, heat and work, while the statements of conservation of energy discussed in Chapter 6 are concerned with kinetic energy and gravitational or elastic potential energy.

14.4 Your initial gravitational potential energy decreases because you ran down the stairs. Your kinetic energy while you are running is positive, but it becomes zero when you stop.

14.7 (a) Heat is transferred out of your body to the colder water.
(b) Your body does work to lift the barbell.
(c) Liposuction decreases your internal energy by removing fat.

14.10 (a) In an isobaric process, heat is not completely converted to work because the internal energy of the system increases.
(b) In an isothermal process, heat is directly converted to work because the internal energy of the system does not change.
(c) In an adiabatic process, heat must be first added to increase the internal energy of the system, and then it is converted to work because in an adiabatic process no heat is transferred.

14.13 In a rapidly expanding gas, there is no heat transfer, so the system is adiabatic. In an adiabatic process, since $Q = 0$, $\Delta U = -W$. The temperature must decrease during an adiabatic process, since work is done at the expense of internal energy.

14.16 An isothermal process in inherently slow, because heat must be added continuously to keep the gas temperature constant at all times and must be allowed to spread through the gas so that there are no hot or cold regions. Isobaric and isochoric processes can occur much faster because keeping the pressure constant or the volume constant does not require dynamic adjustments of the gas.

14.19 Irreversible processes are dissipative and therefore reduce the overall efficiency by converting some of the engine's work output back into heat. A truly reversible process does not lose heat to friction and other dissipative processes and therefore is more efficient.

14.22 The system should have low entropy to produce more work. When entropy increases, a certain amount of energy becomes permanently unavailable to do work. Since entropy is a measure of disorder, the system that is orderly and structured will produce more work.

14.25 A uniform temperature gas has more entropy than one with several different temperatures. The uniform temperature gas is less orderly, less structured, and some energy has become unavailable to do work. The non-uniform temperature gas can convert heat to work without heat transfer from another system by equilibrating its temperature.

14.28 The entropy of a star decreases as it radiates, while the entropy of the space into which it radiates increases. The sum result is that the entropy of the universe increases because the increase in entropy of space is larger than the decrease in entropy for the star.

14.31 When the system equilibrates to 30.0°C , energy has been lost to the system and has become unavailable to do work. The entropy has increased because the decrease in entropy of the hot water is smaller than the increase in entropy of the cold water. The likelihood of the water separating into two equal masses of 20.0°C and 40.0°C again is very small and this likelihood decreases drastically as the mass of the water is increased.

14.34 Refrigerators, air conditioners and heat pumps remove heat from the cold areas and bring it to the hot areas, so the smaller the difference in temperature, the less work that must be done to move the heat. The COP is largest for small temperature differences because the COP is inversely proportional to the difference in temperature. So, the larger the COP, the more cost efficient the system.

PROBLEMS

14.1 Using the energy content of a gallon of gasoline from Table 6.1, the energy stored in the 12.0 gallons of gasoline is: $E_{gas} = (1.3 \times 10^8 \text{ J/gal})(12.0 \text{ gal}) = 1.6 \times 10^9 \text{ J}$. Therefore, the internal energy of the car increases by this energy, so that $\Delta U = \boxed{1.6 \times 10^9 \text{ J}}$

14.7 (a) The metabolic rate is the power, so that: $P = \dfrac{Q}{t} = \dfrac{2500 \text{ kcal}}{1 \text{ day}} \times \dfrac{4186 \text{ J}}{1 \text{ kcal}} \times \dfrac{1 \text{ day}}{8.64 \times 10^4 \text{ s}} = \boxed{121 \text{ W}}$

 (b) Efficiency is defined to be the ratio of what we get to what we spend, or $Eff = \dfrac{W}{E_{in}}$, so we can determine the work done, knowing the efficiency: $W = Eff \cdot E_{in} = 0.200(2500 \text{ kcal}) \times \dfrac{4186 \text{ J}}{1 \text{ kcal}} = \boxed{2.09 \times 10^6 \text{ J}}$

 (c) To compare with a 0.250 hp motor, we need to know how much work the motor does in one day:

 $W = Pt = (0.250 \text{ hp})(1 \text{ day}) \times \dfrac{746 \text{ W}}{1 \text{ hp}} \times \dfrac{8.64 \times 10^4 \text{ s}}{1 \text{ day}} = \boxed{1.61 \times 10^7 \text{ J}}$. So, the man's work output is

 $\dfrac{W_{motor}}{W_{man}} = \dfrac{1.61 \times 10^7 \text{ J}}{2.09 \times 10^6 \text{ J}} = 7.70 \Rightarrow \boxed{7.70 \text{ times less than motor}}$.

14.13 First, we must assume that the volume remains constant, so that $V_1 = V_2$, where state 1 is that at $P_1 = 0.200 \text{ atm} + P_a = 0.200 \text{ atm} + 1.00 \text{ atm} = 1.20 \text{ atm}$ and state 2 is that at $P_2 = 1.00 \text{ atm}$. Now, we can calculate the internal energy of the system in state 2 using Equation 14.3, since helium is a monatomic gas:

$$U_2 = \frac{3}{2} N_2 kT = \frac{3}{2} \left(\frac{P_2 V}{kT} \right) kT = \frac{3}{2} P_2 V = \frac{3}{2} \left(1 \text{ atm} \times \frac{1.013 \times 10^5 \text{ N/m}^2}{1 \text{ atm}} \right) \left(10.0 \text{ L} \times \frac{10^{-3} \text{ m}^3}{1 \text{ L}} \right) = 1.52 \times 10^3 \text{ J}$$

Next, we can use the ideal gas law, in combination with Equation 14.3 to get an expression for U_1 :

$$\frac{U_1}{U_2} = \frac{(3/2) N_1 kT}{(3/2) N_2 kT} = \frac{N_1}{N_2} = \frac{P_1 V / kT}{P_2 V / kT} = \frac{P_1}{P_2} ,$$

so that

$$U_1 = \left(\frac{P_1}{P_2} \right) U_2 = \left(\frac{1.20 \text{ atm}}{1.00 \text{ atm}} \right) (1.52 \times 10^3 \text{ J}) = 1.82 \times 10^3 \text{ J}$$

and so the internal energy inside the balloon is:

$$U_1 - U_2 = 1.82 \times 10^3 \text{ J} - 1.52 \times 10^3 \text{ J} = \boxed{300 \text{ J}} \text{ greater than it would be at zero gauge pressure.}$$

14.19 (a) The efficiency is the work out divided by the heat in:

$$Eff = \frac{W}{Q_h} = \frac{1.50 \times 10^5 \text{ J}}{2.50 \times 10^6 \text{ J}} = 0.0600, \text{ or } \boxed{6.00\%}$$

(b) The work output is the difference between the heat input and the wasted heat, so using Equation 14.4:

$$W = Q_h - Q_c \Rightarrow Q_c = Q_h - W = 2.50 \times 10^6 \text{ J} - 1.50 \times 10^5 \text{ J} = \boxed{2.35 \times 10^6 \text{ J}}.$$

14.25 The maximum efficiency is the Carnot efficiency, so using Equation 14.6, and remembering to convert the temperatures into units of Kelvin, we have:

$$Eff_C = 1 - \frac{T_c}{T_h} = 1 - \frac{278 \text{ K}}{293 \text{ K}} = \boxed{0.0512}, \text{ or } \boxed{5.12\%}$$

14.31 (a)

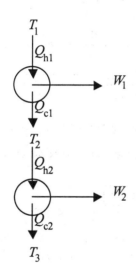

The efficiency is equal to the work out divided by the heat in, so substituting into Equation 14.5, using the diagram, we get:

$$Eff_{total} = \frac{W_1 + W_2}{Q_{h,1}}.$$

But, we know from the individual engines that

$$W_1 = Q_{h,1} - Q_{c,1}, \ W_2 = Q_{h,2} - Q_{c,2}, \text{ and } Q_{c,1} = Q_{h,2},$$

so that

$$W_1 + W_2 = Q_{h,1} - \cancel{Q_{c,1}} + \cancel{Q_{h,2}} - Q_{c,2} = Q_{h,1} - Q_{c,2}.$$

Substituting into our total efficiency formula gives:

$$Eff_{total} = \frac{Q_{h,1} - Q_{c,2}}{Q_{h,1}} = 1 - \frac{Q_{c,2}}{Q_{h,1}}.$$

Finally, since we are dealing with Carnot engines, we can replace the ratio of the heats by the ratio of the absolute temperatures, giving:

$$Eff_{C,total} = 1 - \frac{T_3}{T_1}.$$

In other words, the overall efficiency of two Carnot engines run in series is the same as a single Carnot engine operating between the highest and lowest temperatures, T_1, and T_3.

(b) If real engines are used, the efficiencies would be less than the Carnot efficiencies, and therefore, the two engines would be less efficient than having only one real engine.

14.37 (a) Using Equation 14.8: $COP_{ref} = \dfrac{Q_c}{W} = COP_{hp} - 1$, and for the best coefficient of performance, that means make the Carnot substitution, remembering to use the absolute temperatures:

$$COP_{ref} = \frac{1}{Eff_C} - 1 = \frac{T_h}{T_h - T_c} - 1 = \frac{T_h - (T_h - T_c)}{T_h - T_c} = \frac{T_c}{T_h - T_c} = \frac{243\ \text{K}}{318\ \text{K} - 243\ \text{K}} = \boxed{3.24}$$

(b) Use Equation 14.8, again, and solve for the work done given $Q_c = 1000$ kcal :

$$COP_{ref} = \frac{Q_c}{W} \Rightarrow W = \frac{Q_c}{COP_{ref}} = \frac{1000\ \text{kcal}}{3.24} = 308.6\ \text{kcal} = \boxed{309\ \text{kcal}}$$

(c) The cost is found by converting the units of energy into units of cents:

$$\text{cost} = (308.6\ \text{kcal})\left(\frac{4186\ \text{J}}{1\ \text{kcal}}\right)\left(\frac{10.0\cent}{3.60 \times 10^6\ \text{J}}\right) = \boxed{3.59\cent}$$

(d) We want to determine Q_h, so using Equation 14.4 gives:

$$W = Q_h - Q_c \Rightarrow Q_h = W + Q_c = 309\ \text{kcal} + 1000\ \text{kcal} = \boxed{1309\ \text{kcal}}$$

(e) The inside of this refrigerator (actually freezer) is at $-22.0°\text{F}\ (-30.0°\text{C})$, so this probably is a commercial meatpacking freezer. The exhaust is generally vented to the outside, so as to not heat the building too much.

14.43 (a) Use Equation 14.9 to calculate the change in entropy, remembering to use temperatures in Kelvin:

$$\Delta S = \frac{-Q_h}{T_h} + \frac{Q_c}{T_c} = Q\left(\frac{1}{T_c} - \frac{1}{T_h}\right) = (5.00 \times 10^8\ \text{J})\left(\frac{1}{278\ \text{K}} - \frac{1}{294\ \text{K}}\right) = \boxed{9.79 \times 10^4\ \text{J/K}}$$

(b) In order to gain more energy, we must generate it from things within the house, like a heat pump, human bodies, and other appliances. As you know, we use a lot of energy to keep our houses warm in the winter, because of the loss of heat to the outside.

14.49 When water condenses, it should seem reasonable that its entropy decreases, since the water gets more ordered, so

$$\Delta S = \frac{Q}{T} = \frac{-mL_v}{T} = \frac{-(25.0 \times 10^{-3}\ \text{kg})(2450 \times 10^3\ \text{J/kg})}{308\ \text{K}} = \boxed{-199\ \text{J/K}}\ .$$

The entropy of the water decreases by 199 J/K when it condenses.

14.55 (a) From Table 14.4, we can tabulate the number of ways of getting the three most likely microstates:

H	T	No. of ways
49	51	9.9×10^{28}
50	50	1.0×10^{29}
51	49	9.9×10^{28}

$$\text{Total} = 2.98 \times 10^{29} = \boxed{3.0 \times 10^{29}}$$

(b) The total number of ways is 1.27×10^{30}, so the percent represented by the three most likely microstates is:

$$\% = \frac{\text{total \# of ways to get 3 macrostates}}{\text{total \# of ways}} = \frac{2.98 \times 10^{29}}{1.27 \times 10^{30}} = 0.235 = \boxed{23\%}$$

14.61 (a) The waste heat produced by the engine is Q_c, and since the efficiency is $Eff = \dfrac{W}{Q_h}$, where $W = Q_h - Q_c$, we

can solve for the waste heat, given $W = 2.00 \times 10^8$ J and $Eff = 0.0500$: $W = Eff \times Q_h = Eff\left(W + Q_c\right)$

$$Q_c = \left(\frac{1}{Eff} - 1\right) W = \left(\frac{1}{0.0500} - 1\right)\left(2.00 \times 10^8 \text{ J}\right) = \boxed{3.80 \times 10^9 \text{ J}}.$$

Now, we can solve for the mass of water needed to absorb that heat, using Equation 13.2, $Q = mc\Delta T$, where $\Delta T = 5.00°C$ and $c = 4186$ J/kg·°C :

$$m = \frac{Q}{c\Delta T} = \frac{3.80 \times 10^9 \text{ J}}{\left(4186 \text{ J/kg·°C}\right)\left(5.00°C\right)} = 1.816 \times 10^5 \text{ kg} = \boxed{1.82 \times 10^5 \text{ kg}} \text{ of water.}$$

(b) The flow rate can be found from Equation 11.1, where $t = 2.00$ min $= 120$ s and $\rho = 1000$ kg/m^3 :

$$F = \frac{V}{t} = \frac{m}{\rho t} = \frac{\left(1.816 \times 10^5 \text{ kg}\right)}{\left(1000 \text{ kg/m}^3\right)\left(120 \text{ s}\right)} = \boxed{1.51 \text{ m}^3/\text{s}}.$$

14.67 (a) If it is 0.700 times the Carnot efficiency, we can use Equation 14.6 to determine its efficiency:

$$Eff = 0.700 Eff_C = \frac{0.700\left(T_h - T_c\right)}{T_h} = \frac{0.700\left(293 \text{ K} - 278 \text{ K}\right)}{293 \text{ K}} = 0.0358 = \boxed{3.58\%}.$$

(b) From Equation 14.5a: $Eff = \dfrac{W}{Q_h}$, so that $Q_h = \dfrac{W}{Eff}$, and since $W = Pt$, the rate of heat input is:

$$\frac{Q_h}{t} = \frac{P}{Eff} = \frac{\left(500 \times 10^6 \text{ J/s}\right)}{0.03584} = 1.395 \times 10^{10} \text{ J/s} = \boxed{1.40 \times 10^{10} \text{ J/s or } 3.33 \times 10^6 \text{ kcal/s}}$$

(c) Now, using Equations 13.2 and 11.1, we can determine the flow rate: $Q_h = mc\Delta T = \left(\rho V\right)c\Delta T = \left(\rho F \, t\right)c\Delta T$:

$$F = \frac{Q_h/t}{\rho c\Delta T} = \frac{\left(1.395 \times 10^{10} \text{ J/s}\right)}{\left(1050 \text{ kg/m}^3\right)\left(1.00 \text{ kcal/kg·°C}\right)\left(15.0°C\right)} \times \left(\frac{1 \text{ kcal}}{4186 \text{ J}}\right) = 211.6 \text{ m}^3/\text{s} = \boxed{212 \text{ m}^3/\text{s}}$$

(d) To determine the dollar value, simply convert the units for the 500 MW produced:

$$\$ \text{ value} = \left(500 \times 10^6 \text{ J/s}\right) \times \frac{3.156 \times 10^7 \text{ s}}{1 \text{ y}} \times \frac{10.0¢}{3.60 \times 10^6 \text{ J}} \times \frac{\$1.00}{100¢} = \$4.38 \times 10^8 = \boxed{\$438 \text{ million}}$$

(e) Over ten years, the plant would generate \$4.38 billion, assuming its running at peak capacity at all times. Therefore, construction and operating costs must be less than \$4.38 billion in order for the plant to pay for itself in 10 years. This must not be practical, since you don't see any of these plants around.

15 OSCILLATORY MOTION AND WAVES

CONCEPTUAL QUESTIONS

15.1 The force constant is a measure of the stiffness of a spring. The elastic moduli are also measures of the stiffness of a material, so the force constant is directly proportional to γ, S, and B.

15.4 Simple harmonic motion is oscillatory motion for a system that obeys Hooke's Law. A simple harmonic oscillator will have displacement that goes as Equation 15.5.

15.7 The stiffer the material, the larger the force constant. So according to Equation 15.4b, the stiffer material will have a higher frequency because the frequency is directly proportional to the square root of the force constant.

15.10 The acceleration of a simple harmonic oscillator is zero when its displacement is zero and the velocity is a maximum. The velocity is zero when the displacement is a maximum and the acceleration is a maximum.

15.13 According to Equation 15.8, the length of the pendulum is proportional to the acceleration due to gravity, so if the acceleration of gravity is slightly greater, you will need to slightly lengthen the pendulum to keep the period the same.

15.16 A playground swing is an example of a damped harmonic oscillator. It's amplitude decreases with time because of friction in the hinge and air resistance.

15.19 When you swing on a swing and get yourself to go very high you are in resonance. You move your feet and body (i.e. drive the system) at a frequency that matches the natural frequency of the swing. A young child, on the other hand, works very hard to move a swing, but if he hasn't figured out how to swing his feet and body he will not get the swing to go very high. He hasn't matched his drive frequency with the natural frequency of the swing.

15.22 Dropping a rock in a pond creates transverse waves in the water. The water moves up and down while the wave propagates across the pond, perpendicular to the motion of the water. Shock waves are compression waves, where the air is compressed in the direction of the motion of the wave.

15.25 If the speakers are connected in opposite ways, then the sound waves will produce destructive interference when they should be producing constructive interference, and the music will not sound right. If they are both connected the wrong way (or the right way) the sound will be okay.

15.28 The energy and intensity of a wave are proportional to the average elastic potential energy over one cycle. The intensity would not necessarily be proportional to amplitude squared if the medium did not follow Hooke's Law. If Hooke's Law was violated, then the work, which is proportional to force times distance, would not be proportional to distance squared, since the force would not necessarily be proportional to distance displaced.

PROBLEMS

15.1 Using Equation 15.1, $F = -kx$, where $m = 120$ kg and $x = -0.750 \times 10^{-2}$ m, gives:

$$k = -\frac{F}{x} = -\frac{mg}{x} = -\frac{(120 \text{ kg})(9.80 \text{ m/s}^2)}{-0.750 \times 10^{-2} \text{ m}} = \boxed{1.57 \times 10^5 \text{ N/m}}.$$

The force constant must be a positive number!

15.7 (a) Using Equation 15.1, $F = -kx$, where $m = 80.0$ kg and $x = -0.120$ m, gives:

$$k = -\frac{F}{x} = -\frac{mg}{x} = -\frac{(80.0 \text{ kg})(9.80 \text{ m/s}^2)}{-0.120 \text{ m}} = \boxed{6.53 \times 10^3 \text{ N/m}}$$

(b) Yes, when the man is at his lowest point in his hopping the spring will be compressed the most.

15.13 Using Equation 15.3, where $f = 60.0$ Hz $= 60.0$ s^{-1}, gives:

$$T = \frac{1}{f} = \frac{1}{60.0 \text{ Hz}} = \boxed{1.67 \times 10^{-2} \text{ s}}$$

15.19 (a) Using Equation 15.4a: $T = 2\pi\sqrt{\dfrac{m}{k}}$, where $m = 4.00 \times 10^5$ kg and $T = 2.00$ s, gives:

$$k = \frac{4\pi^2 m}{T^2} = \frac{4\pi^2 (4.00 \times 10^5 \text{ kg})}{(2.00 \text{ s})^2} = \boxed{3.95 \times 10^6 \text{ N/m}}$$

(b) Using Equation 15.2, where $x = 2.00$ m, gives:

$$PE_{el} = \frac{1}{2}kx^2 = \frac{1}{2}(3.95 \times 10^6 \text{ N/m})(2.00 \text{ m})^2 = \boxed{7.90 \times 10^6 \text{ J}}$$

15.25 Using Equation 15.4a, for each mass: $T_1 = 2\pi\sqrt{\dfrac{m_1}{k}}$; $T_2 = 2\pi\sqrt{\dfrac{m_2}{k}}$, so the ratio of the periods can be written in terms of their masses:

$$\frac{T_1}{T_2} = \sqrt{\frac{m_1}{m_2}} \Rightarrow m_2 = \left(\frac{T_2}{T_1}\right)^2 m_1 = \left(\frac{2.00 \text{ s}}{1.50 \text{ s}}\right)^2 (0.500 \text{ kg}) = 0.889 \text{ kg}.$$

The final mass minus the initial mass gives the mass that must be added:

$$\Delta m = m_2 - m_1 = 0.889 \text{ kg} - 0.500 \text{ kg} = \boxed{0.389 \text{ kg}}.$$

15.31 Use Equation 15.8, where $L = 1.00$ m :

$$T = 2\pi\sqrt{\frac{L}{g}} = 2\pi\sqrt{\frac{1.00 \text{ m}}{9.80 \text{ m/s}^2}} = \boxed{2.01 \text{ s}}$$

15.37 (a) Use Equation 15.8, and write the new period in terms of the old period, where $L' = 2L$:

$$T' = 2\pi\sqrt{\frac{2L}{g}} = \sqrt{2}\left(2\pi\sqrt{\frac{L}{g}}\right) = \sqrt{2}\ T.$$

The period $\boxed{\text{increases by a factor of } \sqrt{2}.}$

(b) This time, $L' = 0.950L$, so that

$$T' = 2\pi\sqrt{\frac{0.950\ L}{g}} = \sqrt{0.950}\ T.$$

The new period is decreased by a factor of $\sqrt{0.950}$ or it is $\boxed{97.5\%}$ of the old period.

15.43 Use Equation 15.6, since the person bounces in simple harmonic motion, where $X = 0.200 \times 10^{-2}$ m:

$$v_{max} = \sqrt{\frac{k}{m}}X = \sqrt{\frac{1.50 \times 10^5\ \text{N/m}}{55.0\ \text{kg}}}\left(0.200 \times 10^{-2}\ \text{m}\right) = \boxed{0.104\ \text{m/s}}$$

15.49 Recall Equation 9.2, $v = R\omega$, so that

$$\omega = \frac{v}{R} = \frac{40.0\ \text{m/s}}{0.600\ \text{m}} = 66.67\ \text{rad/s}.$$

The velocity of the piston is then found using Equation 9.2, where $r = 0.450$ m:

$$v_p = r\omega = (0.450\ \text{m})(66.67\ \text{rad/s}) = 30.0\ \text{m/s}$$

(relative to locomotive.).

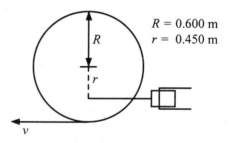

R = 0.600 m
r = 0.450 m

Since the piston moves in simple harmonic motion, the velocity we found above is actually the maximum velocity: $v_{max,\,p} = \boxed{30.0\ \text{m/s}}$.

15.55 (a) We know that there are two forces acting on the mass in the horizontal direction: net $F = \mu_s N - kx = 0$, where x is the distance the spring is stretched. Given: $\mu_s = 0.100$, $m = 0.750$ kg, $k = 150$ N/m, and since the mass is on a horizontal surface, $N = mg$, so we can solve for x:

$$x = \frac{\mu_s mg}{k} = \frac{(0.100)(0.750\ \text{kg})(9.80\ \text{m/s}^2)}{150\ \text{N/m}} = \boxed{4.90 \times 10^{-3}\ \text{m}}.$$

So, the maximum distance the spring can be stretched without moving the mass is 4.90×10^3 m.

(b) From Example 15.7, given $X = 2x$, $k = 150$ N/m, $\mu_k = 0.0850$, and $m = 0.750$ kg:

$$d = \frac{(1/2)k(2x)^2}{\mu_k mg} = \frac{0.5(150\ \text{N/m})(9.80 \times 10^{-3}\ \text{m})^2}{(0.0850)(0.750\ \text{kg})(9.80\ \text{m/s}^2)} = \boxed{1.15 \times 10^{-2}\ \text{m}},$$

which is the total distance traveled by the mass before it stops.

15.61 Use Equation 15.11, $v_w = f\lambda$, where $v_w = 5.00$ m/s and $\lambda = 40.0$ m :

$$f = \frac{v_w}{\lambda} = \frac{5.00 \text{ m/s}}{40.0 \text{ m}} = 0.125 \text{ Hz}$$

Now that we know the frequency, we can calculate the number of oscillations:

$$N = ft = (0.125 \text{ Hz})(60.0 \text{ s}) = \boxed{7.50} \ .$$

15.67 Use the definition of velocity, $v = \frac{d}{t}$, given the wave velocity and the time:

$$d = v_w t = (340 \text{ m/s})(1.00 \times 10^{-3} \text{ s}) = 0.340 \text{ m} = \boxed{34.0 \text{ cm}} \ .$$

Therefore, objects that are 34.0 cm apart (or farther) will produce sounds that the human ear can distinguish.

15.73 There will be three different beat frequencies because of the interactions between the three different frequencies. Using Equation 15.12 gives:

$$f_{B1} = |370 \text{ Hz} - 349 \text{ Hz}| = \boxed{21 \text{ Hz}} \ ,$$

$$f_{B2} = |392 \text{ Hz} - 370 \text{ Hz}| = \boxed{22 \text{ Hz}} \ , \text{ and}$$

$$f_{B3} = |392 \text{ Hz} - 349 \text{ Hz}| = \boxed{43 \text{ Hz}}$$

15.79 First, we know, from Equation 15.13: $I = \frac{P}{A}$, where the power can be expressed using Equation 6.9: $P = \frac{W}{t}$, and

the area is the surface area of a circle: $A = \pi r^2$. Combining these equations gives:

$$I = \frac{W}{t(\pi r^2)}$$

Then, solving for the time gives:

$$t = \frac{W}{\pi r^2 I} = \frac{(1.00 \text{ kcal})(4186 \text{ J/kcal})}{\pi (0.0300 \text{ m})^2 (5.00 \times 10^3 \text{ W/m}^2)} = 296 \text{ s} = \boxed{4.94 \text{ min}}$$

15.85 (a) Using Equations 15.13 and 6.9, we see that

$$W = Pt = (0.900 IA)t = 0.900 I \pi r^2 t \ ,$$

so that:

$$I = \frac{W}{0.900 \pi r^2 t} = \frac{500 \text{ J}}{0.900 \pi (1.00 \times 10^{-3} \text{ m})^2 (4.00 \text{ s})} = \boxed{4.42 \times 10^7 \text{ W/m}^2}$$

(b) The intensity of a laser is about 4×10^4 times that of the sun, so clearly lasers can be very damaging if they enter your eye! This means that staring into a laser for one second is equivalent to staring at the sun for 11 hours without blinking!!

15.91 (a) Examine Chapter 3 to review how to solve projectile motion problems.
Given: $x = 20.0 \text{ m}$, $y = -10.0 \text{ m}$, and $\theta = 42°$. Find: v_0. We know

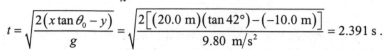

$$x = (v_0 \cos\theta_0)t, \quad y = (v_0 \sin\theta_0)t - \frac{1}{2}gt^2,$$

so that $\tan\theta_0 = \dfrac{y + (1/2)gt^2}{x}$, and solving for time:

$$t = \sqrt{\frac{2(x\tan\theta_0 - y)}{g}} = \sqrt{\frac{2\left[(20.0 \text{ m})(\tan 42°) - (-10.0 \text{ m})\right]}{9.80 \text{ m/s}^2}} = 2.391 \text{ s}.$$

Now, substituting into the horizontal motion equation, we can solve for v_0:

$$v_0 = \frac{x}{(\cos\theta_0)t} = \frac{20.0 \text{ m}}{(\cos 42°)(2.391 \text{ s})} = 11.26 \text{ m/s} = \boxed{11.3 \text{ m/s}}$$

(b) Use conservation of energy:

$$\frac{1}{2}mv^2 + mgh = \frac{1}{2}kx^2;$$

where $h = x\sin 42°$, and $x = 15.0 \text{ m}$:

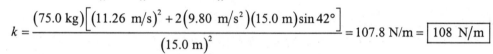

$$k = \frac{mv^2}{x^2} + \frac{2mgh}{x^2} = \frac{m(v^2 + 2gx\sin\theta)}{x^2}, \text{ so that}$$

$$k = \frac{(75.0 \text{ kg})\left[(11.26 \text{ m/s})^2 + 2(9.80 \text{ m/s}^2)(15.0 \text{ m})\sin 42°\right]}{(15.0 \text{ m})^2} = 107.8 \text{ N/m} = \boxed{108 \text{ N/m}}.$$

(c) The force produced by the cannon's spring is given by Equation 15.1:

$$F = ma = kx \Rightarrow a = \frac{kx}{m} = \frac{(107.8 \text{ N/m})(15.0 \text{ m})}{75.0 \text{ kg}} = 21.56 \text{ m/s}^2 \times \frac{g}{9.80 \text{ m/s}^2} = \boxed{2.20g}$$

Yes, a person can withstand this amount of acceleration. For example, the driver of the rocket sled in Example 4.2 would be accelerated at close to 5g's.

SOUND AND HEARING (16)

CONCEPTUAL QUESTIONS

16.1 Sound is a disturbance of matter that is transmitted from its source outward and is one type of wave. On the atomic scale, it is a disturbance of atoms that is far more ordered than their thermal motions.

16.4 The speed of sound in a medium is determined by a combination of its rigidity and its density. We would expect the speed of sound in marble to be faster because it is more rigid (γ is larger, see Table 5.1) and it is less dense than lead ($\rho_{marble} = 2.7 \times 10^3$ kg/m^3 and $\rho_{lead} = 11.3 \times 10^3$ kg/m^3).

16.7 The Doppler shift is real; it is a change in the frequency of the sound due to motion.

16.10 Observers on the ground often do not see the aircraft creating the sonic boom, because it has passed by before the shock wave reaches them.

16.13 The intensity of sound from an unamplified guitar is larger than that from a string held taut by a simple stick because the guitar uses resonance in its sounding box to amplify and enrich the sound created by its vibrating strings.

16.16 Overtones are all higher resonant frequencies than the fundamental, while harmonics are all resonant frequencies. Therefore, all overtones are harmonics, but not all harmonics are overtones.

16.19 Figure 16.22 shows the relationship of loudness in phons to intensity in decibels for persons of normal hearing. Your hearing may either be better or worse than the normal person, so it is possible for your threshold of hearing to be 0 dB at 250 Hz.

16.22 The physical characteristics associated with loudness are intensity and frequency. Loudness is not a physical characteristic of sound, but is the perception of intensity.

16.25 The advantage of infrasound is that it is low frequency (less than 20 Hz), and therefore is not absorbed very strongly. The higher the frequency the more the absorption and the shorter the distance the sound can travel. Also, since elephants can hear at such low frequencies, they can communicate with each other without being heard by people and other animals.

16.28 The high frequencies used in ultrasonic cleaners prevent it from penetrating too deeply into the object, since the higher the frequency the more strongly it is absorbed. The intensity is great enough to cause cavitation, which is responsible for most of the cleansing action. The shock pressures created by cavitation reach into small crevices where even a low surface tension cleaning fluid might not penetrate.

16.31 The density of the medium transmitting the sound wave changes drastically from the loudspeaker to the air, so the sound waves tend to be reflected back into the speaker rather than being transmitted into the air.

PROBLEMS

16.1   Use Equation 16.1, $v_w = f\lambda$, where $f = 1200$ Hz and $v_w = 345$ m/s :

$$\lambda = \frac{v_w}{f} = \frac{345 \text{ m/s}}{1200 \text{ Hz}} = \boxed{0.288 \text{ m}}$$

16.7   The wavelength of sounds in air and water are different because the speed of sound is different in air and water. We know $v_{\text{seawater}} = 1540$ m/s (from Table 16.1) and $v_{\text{air}} = 343$ m/s, at 20.0°C, from Problem 16.5, so from Equation 16.1 we know: $v_{\text{seawater}} = f\lambda_{\text{seawater}}$ and $v_{\text{air}} = f\lambda_{\text{air}}$, so we can determine the ratio of the wavelengths:

$$\frac{v_{\text{air}}}{v_{\text{seawater}}} = \frac{\lambda_{\text{air}}}{\lambda_{\text{seawater}}} \Rightarrow \frac{\lambda_{\text{air}}}{\lambda_{\text{seawater}}} = \frac{343 \text{ m/s}}{1540 \text{ m/s}} = \boxed{0.223}$$

16.13 From Equation 16.4, we know: $\beta = 10\log_{10}\left(\dfrac{I}{I_0}\right)$, where $I_0 = 10^{-12}$ W/m² so that $I = I_0 10^{\beta/10}$. For $\beta = 91.0$ dB,

$$I = \left(1.00 \times 10^{-12} \text{ W/m}^2\right)10^{91.0/10.0} = \boxed{1.26 \times 10^{-3} \text{ W/m}^2}.$$

(To calculate an exponent that is not an integer, use the x^y-key on your calculator.)

16.19 Use Equation 16.4, $\beta = 10\log\left(\dfrac{I}{I_0}\right)$, where $I_0 = 10^{-12}$ W/m² and $\beta = -8.00$ dB, so that:

$$I = I_0 10^{\beta/10} = \left(1.00 \times 10^{-12} \text{ W/m}^2\right)10^{-8.00/10.0} = \boxed{1.58 \times 10^{-13} \text{ W/m}^2}$$

16.25 (a)   Using Equation 16.3, $I = \dfrac{P}{A}$, we see that for the same power, $\dfrac{I_2}{I_1} = \dfrac{A_1}{A_2}$, so for a 5.00% efficiency:

$$\frac{I_e}{I_t} = \frac{A_t}{A_e} = \frac{(0.0500)\left(900 \text{ cm}^2\right)}{0.500 \text{ cm}^2} = 90 .$$

Now, using Equation 16.4, and remembering that $\log A - \log B = \log\dfrac{A}{B}$, we see that:

$$\beta_e - \beta_t = 10\log\left(\frac{I_e}{I_0}\right) - 10\log\left(\frac{I_t}{I_0}\right) = 10\log\left(\frac{I_e}{I_t}\right) = 10\log(90) = 19.54 \text{ dB} = \boxed{19.5 \text{ dB}}$$

(b)   This increase of approximately 20 dB, increases the sound of a normal conversation to that of a loud radio or classroom lecture, see Table 16.2. For someone who cannot hear at all, this will not be helpful, but for someone who is starting to lose their ability to hear, it may help. Unfortunately, ear trumpets are very cumbersome, so even though they could help, the inconvenience makes them rather impractical.

16.31 We can use Equation 16.5 (with the minus sign because the source is approaching), $f_{obs} = f_s \dfrac{v_w}{v_w - v_s}$, to determine

the speed of the musician (the source), given $f_{obs} = 888$ Hz, $f_s = 880$ Hz, and $v_w = 338$ m/s:

$$v_s = \frac{v_w \left(f_{obs} - f_s\right)}{f_{obs}} = \frac{(338 \text{ m/s})(888 \text{ Hz} - 880 \text{ Hz})}{888 \text{ Hz}} = \boxed{3.05 \text{ m/s}}$$

16.37 (a) Using Equation 15.12: $f_{B, A\&C} = \left|f_1 - f_2\right| = |264 \text{ Hz} - 220 \text{ Hz}| = \boxed{44 \text{ Hz}}$

(b) $f_{B, D\&F} = \left|f_1 - f_2\right| = |352 \text{ Hz} - 297 \text{ Hz}| = \boxed{55 \text{ Hz}}$

(c) We get beats from every combination of frequencies, so in addition to the two beats above, we also have:

$$f_{B, F\&A} = 352 \text{ Hz} - 220 \text{ Hz} = \boxed{132 \text{ Hz}}, \; f_{B, F\&C} = 352 \text{ Hz} - 264 \text{ Hz} = \boxed{88 \text{ Hz}},$$

$$f_{B, D\&C} = 297 \text{ Hz} - 264 \text{ Hz} = \boxed{33 \text{ Hz}}, \text{ and } f_{B, D\&A} = 297 \text{ Hz} - 220 \text{ Hz} = \boxed{77 \text{ Hz}}$$

16.43 We know from Equation 16.8, that the frequency for a tube open at both ends is: $f_n = n\left(\dfrac{v}{2L}\right)$ for $n = 1, 2, 3 \dots$.

If the fundamental frequency ($n = 1$) is: $f_1 = 262$ Hz, we can determine the length: $f_1 = \dfrac{v_w}{2L} \Rightarrow L = \dfrac{v_w}{2f_1}$. We

need to determine the speed of sound, from Equation 16.2, since we are told the air temperature:

$$v_w = \left(331 \text{ m/s}\right)\sqrt{\frac{T(\text{K})}{273 \text{ K}}} = \left(331 \text{ m/s}\right)\sqrt{\frac{293 \text{ K}}{273 \text{ K}}} = 342.9 \text{ m/s}.$$

So,

$$L = \frac{342.9 \text{ m/s}}{2\left(262 \text{ Hz}\right)} = 0.654 \text{ m} = \boxed{65.4 \text{ cm}}$$

16.49 First, we need to determine the speed of sound at $37.0°\text{C}$, using Equation 16.2:

$$v_w = \left(331 \text{ m/s}\right)\sqrt{\frac{T(\text{K})}{273 \text{ K}}} = \left(331 \text{ m/s}\right)\sqrt{\frac{310 \text{ K}}{273 \text{ K}}} = 352.7 \text{ m/s}.$$

Next, using Equation 16.7 for tubes closed at one end: $f_n = n\left(\dfrac{v_w}{4L}\right)$, $n = 1, 3, 5, \dots$, we can determine the

frequency of the first overtone ($n = 3$): $f_3 = 3\dfrac{352.7 \text{ m/s}}{4\left(0.0240 \text{ m}\right)} = 1.10 \times 10^4 \text{ Hz} = \boxed{11.0 \text{ kHz}}$.

The ear is not particularly sensitive to this frequency, so we don't hear overtones due to the ear canal.

16.55 From page 419, we know that we can discriminate between two sounds if their frequencies differ by at least 0.3%, so the closest frequencies to 500 Hz that we can distinguish are

$$f = (500 \text{ Hz})(1 \pm 0.003) = \boxed{498.5 \text{ Hz}} \text{ and } \boxed{501.5 \text{ Hz}}.$$

16.61 Reading from Figure 16.22: a 600 Hz tone at a loudness of 20 phons has a sound level of about $\boxed{23 \text{ dB}}$, while a 600 Hz tone at a loudness of 70 phons has a sound level of about $\boxed{70 \text{ dB}}$.

16.67 From Figure 16.22, the 0 phons line is normal hearing. So, this person can barely hear a 100 Hz sound at 10 dB above normal, requiring a 47 dB sound level (β_1). For a 4000 Hz sound, this person requires 50 dB above normal, or a 43 dB sound level (β_2) to be audible. So, the 100 Hz tone must be 4 dB higher than the 4000 Hz sound. To calculate the difference in intensity, use Equation 16.4 and convert the difference in decibels to a ratio of intensities. $\beta_1 - \beta_2 = 10\log\left(\dfrac{I_1}{I_0}\right) - 10\log\left(\dfrac{I_2}{I_0}\right) = 10\log\left(\dfrac{I_1}{I_2}\right)$ substituting in the values from above gives:

$$10\log\left(\dfrac{I_1}{I_2}\right) = 47 \text{ dB} - 43 \text{ dB} = 4 \text{ dB, or } \dfrac{I_1}{I_2} = 10^{4/10} = \boxed{2.5}.$$

So the 100 Hz tone must be 2.5 times more intense than the 4000 Hz sound to be audible by this person.

16.73 (a) From Table 16.1, the speed of sound in tissue is $v_w = 1540$ m/s, so using Equation 16.1, $v_w = f\lambda$, we find the minimum frequency to resolve 0.250 mm details is:

$$f = \frac{v_w}{\lambda} = \frac{1540 \text{ m/s}}{0.250 \times 10^{-3} \text{ m}} = \boxed{6.16 \times 10^6 \text{ Hz}}.$$

(b) From page 425, we know that the accepted rule of thumb is that you can effectively scan to a depth of about 500λ into tissue, so the effective scan depth is:

$$500\lambda = 500\left(0.250 \times 10^{-3} \text{ m}\right) = 0.125 \text{ m} = \boxed{12.5 \text{ cm}}$$

16.79 This problem requires two steps: (1) determine the frequency the blood receives (which is the frequency that is reflected), then (2) determine the frequency that the scanner receives. So first, the blood is a moving observer, and Equation 16.6 gives the frequency it receives (with the plus sign used because the blood is approaching):

$$f_b = f_s\left(\frac{v_w + v_b}{v_w}\right) \left(\text{where } v_b = \text{blood velocity}\right)$$

Next, this frequency is reflected from the blood, which now acts as a moving source. Equation 16.5 (with the minus sign used because the blood is still approaching) gives the frequency received by the scanner:

$$f'_{obs} = f_b\left(\frac{v_w}{v_w - v_b}\right) = f_s\left(\frac{v_w + v_b}{v_w}\right)\left(\frac{v_w}{v_w - v_b}\right) = f_s\left(\frac{v_w + v_b}{v_w - v_b}\right).$$

Solving for the speed of the blood gives:

$$v_b = v_w\left(\frac{f'_{obs} - f_s}{f'_{obs} + f_s}\right) = \frac{(1540 \text{ m/s})(500 \text{ Hz})}{\left(2.00 \times 10^6 \text{ Hz} + 500 \text{ Hz}\right) + 2.00 \times 10^6 \text{ Hz}} = \boxed{0.192 \text{ m/s}}.$$

The blood's speed is 19.2 cm/s.

16.85 (a) From Equation 6.9, we know $P = \dfrac{E}{t}$, and from Equation 15,13, we know $I = \dfrac{P}{A}$, so that

$$E = Pt = IAt = I\left(\pi r^2\right)t, \text{ where we are given } I = 1.00 \times 10^4 \text{ W/m}^2, r = \frac{0.0350 \text{ m}}{2} = 0.0175 \text{ m, and}$$

$t = 5.00 \text{ min} = 300 \text{ s}$, so that

$$E = \left(1.00 \times 10^4 \text{ W/m}^2\right)\pi\left(1.75 \times 10^{-2} \text{ m}\right)^2(300 \text{ s}) = \boxed{2.89 \times 10^3 \text{ J}}.$$

(b) Using Equation 13.2, $Q = mc\Delta T$, where $m = 0.300$ kg, and $c = 3500$ J/kg·°C (from Table 13.1), we have:

$$\Delta T = \frac{Q}{mc} = \frac{2.89 \times 10^3 \text{ J}}{(0.300 \text{ kg})(3500 \text{ J/kg·°C})} = \boxed{2.75°C}.$$

ELECTRIC CHARGE AND ELECTRIC FIELD

17

CONCEPTUAL QUESTIONS

17.1 Most objects do not exhibit static electricity because there are equal numbers of positive and negative charges present, so the net charge is zero.

17.4 Your finger acts like a dielectric inserted into a capacitor. Polarization of your finger reduces the electric field felt by the leaves causing them to drop. Removing your finger restores the original condition and the leaves rise again.

17.7 Rubbing (or buffing) the car has left it statically charged, just like the "Swiffer" cleaning brooms. This static charge attracts dust even though the car wax and tires are insulators.

17.10 In (c), the negative charges are attracted to the rod and the positive charges are repelled, causing extra distance between the induced charges. In (d), the rod is removed.

17.13 Because water is a polar molecule, it can orient itself so that its negatively charged part is toward the positively charged ion and therefore be attracted. This attraction is what allows water to condense, forming rain droplets.

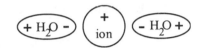

17.16

Coulomb Force Field	Electric Field
• Lines cannot cross.	• Lines cannot cross.
• Direction depends on charge of second particle.	• Direction independent of "test charge".
• Magnitude of force (density of lines) depends on second particle.	• Magnitude (density of lines) independent of "test charge".
• Lines are continuous.	• Lines are continuous.
• Lines can emerge or terminate on any charge. In other words, they can emerge from both positive and negative charges.	• Lines emerge from positive charges and end on negative charges only.

17.19 Yes, because these are very non-uniform field lines, indicating the object has made a significant impact. Insulators can have non-perpendicular electric field lines because they cannot move their charges too far, so an insulator would be unable to move the charges enough to make the field lines perpendicular in this case.

17.22 The golf club, being a conductor, reduces the gap the lightning must jump between the cloud and the ground. A tree might help if she's not too close to it, otherwise the lightning from the tree strike would still produce a potential drop across her, which would be bad.

17.25 A grounded rod would be more likely to be hit, but it would be better able to dissipate the impact of the hit. A rod attached to the building would dissipate the impact of the lightning through the building, which could cause serious power surges.

17.28 (a) The net Coulomb force on q would be to the right.

(b) The direction of the electric field at the center of the square is then to the left because it is the direction of the force on a *positive* test charge.

PROBLEMS

17.1 (a) Since one electron has a charge of $q_e = -1.60 \times 10^{-19}$ C, we can determine the number of electrons necessary to generate a total charge of -2.00 nC by using the equation: $Nq_e = Q$, so that:

$$N = \frac{Q}{q_e} = \frac{-2.00 \times 10^{-9} \text{ C}}{-1.60 \times 10^{-19} \text{ C}} = \boxed{1.25 \times 10^{10} \text{ electrons}}$$

   (b) Similarly, we can determine the number of electrons removed from a neutral object to leave a charge of $0.500 \ \mu$C, because that tells us we had to remove $-0.500 \ \mu$C of charge, so that:

$$N = \frac{-0.500 \times 10^{-6} \text{ C}}{-1.60 \times 10^{-19} \text{ C}} = \boxed{3.13 \times 10^{12} \text{ electrons}}$$

17.7 Recall from Chapter 12, page 303, that Avogadro's number is $N_A = 6.02 \times 10^{23}$ atoms/mole. Now, we need to determine the number of moles of copper that are present. We do this using the mass and the atomic mass:

$$n = \frac{m}{A} = \frac{50.0 \text{ g}}{63.5 \text{ g/mol}} \ .$$

So, since there are 29 protons per atom, we can determine the number of protons, N_p, from:

$$N_p = nN_A \times 29 \text{ protons/atom} = \left(\frac{50.0 \text{ g of Cu}}{63.5 \text{ g/mol}} \right) \left(\frac{6.02 \times 10^{23} \text{ atoms}}{\text{mol}} \right) \times \frac{29 \text{ protons}}{\text{Cu atom}} = 1.375 \times 10^{25} \text{ protons} \ .$$

Since there are the same numbers of electrons as protons in a neutral atom, before we remove the electrons to give the copper a net charge, we have 1.375×10^{25} electrons.

Next, we need to determine the number of electrons we removed to leave a net charge of $2.00 \ \mu$C. We need to remove $-2.00 \ \mu$C of charge, so the number of electrons to be removed is given by:

$$N_{e, \text{ removed}} = \frac{Q}{q_e} = \frac{-2.00 \times 10^{-6} \text{ C}}{-1.60 \times 10^{-19} \text{ C}} = 1.25 \times 10^{13} \text{ electrons removed} \ .$$

Finally, we can calculate the fraction of the copper's electrons by taking the ratio of the number of electrons removed to the number of electrons originally present:

$$\frac{N_{e, \text{ removed}}}{N_{e, \text{ initially}}} = \frac{1.25 \times 10^{13}}{1.375 \times 10^{25}} = \boxed{9.09 \times 10^{-13}}$$

17.13 Using Equation 17.2, we see that the force is inversely proportional to the separation distance squared, so that:

$$F_1 = k \frac{q_1 q_2}{r_1^2} \text{ and } F_2 = k \frac{q_1 q_2}{r_2^2} \ .$$

Since we know the ratio of the forces, we can determine the ratio of the separation distances:

$$\frac{F_1}{F_2} = \left(\frac{r_2}{r_1} \right)^2 \text{ so that } \frac{r_2}{r_1} = \sqrt{\frac{F_1}{F_2}} = \sqrt{\frac{1}{25}} = \frac{1}{5} \ .$$

The separation decreased by a $\boxed{\text{factor of 5}}$.

17.19 (a) If the electrostatic force is to support the weight of a 10.0 mg piece of tape, it must be a force equal to the gravitational force acting on the tape, so using Equation 17.2 and the assumption that the point charges are equal, we can set the electrostatic force equal to the gravitational force.

$$F = k\frac{q_1 q_2}{r^2} = \frac{kq^2}{r^2} = mg \Rightarrow q = \left(\frac{r^2 mg}{k}\right)^{1/2} = \left[\frac{(0.0100 \text{ m})^2 (10.0\times10^{-6} \text{ kg})(9.80 \text{ m/s}^2)}{(9.00\times10^9 \text{ N}\cdot\text{m}^2/\text{C}^2)}\right]^{1/2} = \boxed{1.04\times10^{-9} \text{ C}}$$

(b) This charge is approximately 1 nC, which is consistent with the magnitude of charge typical static electricity.

17.25 (a)

We know that since the negative charge is smaller, the third charge should be placed to the right of the negative charge, if the net force on it to be zero. So, if we want $F_{net} = F_1 + F_2 = 0$,

we can use Equation 17.2 to write the forces in terms of the distances:

$$\frac{kq_1 q}{r_1^2} + \frac{kq_2 q}{r_2^2} = kq\left(\frac{q_1}{r_1^2} + \frac{q_2}{r_2^2}\right) = 0,$$

or since $r_1 = 0.250 \text{ m} + d$ and $r_2 = d$:

$$\frac{5 \text{ }\mu\text{C}}{(0.250 \text{ m} + d)^2} = \frac{3 \text{ }\mu\text{C}}{d^2}, \text{ or } d = \sqrt{\frac{3}{5}}(0.250 \text{ m}) + \sqrt{\frac{3}{5}}d,$$

so that

$$d\left(1 - \sqrt{\frac{3}{5}}\right) = \frac{\sqrt{3}}{5}(0.250 \text{ m}), \text{ and finally, } d = \frac{\sqrt{\frac{3}{5}}(0.250 \text{ m})}{1 - \sqrt{\frac{3}{5}}} = \boxed{0.859 \text{ m}}.$$

The charge must be placed a distance of 0.859 m to the far side of the negative charge.

(b)

This time we know that the charge must be placed between the two positive charges, and closer to the 3 μC charge for the net force to be zero. So, if we want $F_{net} = F_1 + F_2 = 0$, we can again use Equation 17.2 to write the forces in terms of the distances:

$$\frac{kq_1 q}{r_1^2} - \frac{kq_2 q}{r_2^2} = kq\left(\frac{q_1}{r_1^2} - \frac{q_2}{r_2^2}\right) = 0,$$

or since $r_1 = 0.250 \text{ m} - r_2$,

$$\frac{5 \text{ }\mu\text{C}}{(0.250 - r_2)^2} = \frac{3 \text{ }\mu\text{C}}{r_2^2}, \text{ or } r_2^2 = \frac{3}{5}(0.250 \text{ m} - r_2)^2,$$

so that

$$r_2 = \sqrt{\frac{3}{5}}(0.250 \text{ m} - r_2), \text{ or } r_2\left(1 + \sqrt{\frac{3}{5}}\right) = \sqrt{\frac{3}{5}}(0.250 \text{ m}),$$

and finally,

$$r_2 = \frac{\sqrt{\frac{3}{5}}(0.250 \text{ m})}{1 + \sqrt{\frac{3}{5}}} = \boxed{0.109 \text{ m}}.$$

The charge must be placed between the two charges and a distance of 0.109 m from the 3 μC charge.

17.31 (a) Using Equation 17.3, we can find the electric field caused by a given force on a given charge (taking the eastward direction to be positive):

$$E = \frac{F}{q} = \frac{-4.80 \times 10^{-17} \text{ N}}{-1.60 \times 10^{-19} \text{ C}} = \boxed{300 \text{ N/C (east)}}$$

(b) The force should be equal to the force on the electron, only in the opposite direction. Using Equation 17.3, we get:

$$F = qE = \left(1.60 \times 10^{-19} \text{ C}\right)\left(300 \text{ N/C}\right) = \boxed{4.80 \times 10^{-17} \text{ N (east)}},$$

as we expected.

17.37

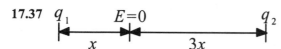

If the electric field is zero ¼ of the way from q_1 to q_2, then we know from Equation 17.4 that:

$$|E_1| = |E_2| \Rightarrow \frac{kq_1}{x^2} = \frac{kq_2}{(3x)^2}, \text{ so that } \frac{q_2}{q_1} = \frac{(3x)^2}{x^2} = \boxed{9}.$$

The charge q_2 is 9 times larger than q_1.

17.43

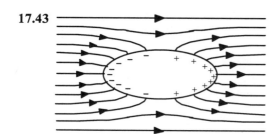

The field lines deviate from their originally horizontal direction because the charges within the object rearrange. The field lines will come into the object perpendicular to the surface, and will leave the other side of the object perpendicular to the surface.

Yes, the field is smaller near the long side of the object. This is evident because there are fewer field lines near the long side of the object, and there are more field lines near the points of the object.

17.49 (a) According to Figure 17.40(b), the point charges are given by: $q_1 = -2.00 \ \mu\text{C}$ at $x = 1.00$ cm ; $q_5 = +1.00 \ \mu\text{C}$ at $x = 5.00$ cm ; $q_8 = +3.00 \ \mu\text{C}$ at $x = 8.00$ cm; and $q_{14} = -1.00 \ \mu\text{C}$ at $x = 14.0$ cm. If a 2.00 nC charge is placed at $x = 4.00$ cm, the force it feels from the other charges is found by using Equation 17.2. The net force is the vector addition of the force due to each point charge, but since the point charges are all along the x-axis, the forces add like numbers. Thus, the net force is given by:

$$F = \frac{kq_1 q}{r_1^2} - \frac{kq_5 q}{r_5^2} - \frac{kq_8 q}{r_8^2} - \frac{kq_{14} q}{r_{14}^2} = kq \left(\frac{q_1}{r_1^2} - \frac{q_5}{r_5^2} - \frac{q_8}{r_8^2} - \frac{q_{14}}{r_{14}^2} \right),$$

to the right. Notice that the term involving the charge q_1 has an opposite sign because it pulls in the opposite direction than the other three charges. Substituting in the values gives:

$$F = \left(\frac{9.00 \times 10^9 \text{ N} \cdot \text{m}^2}{\text{C}^2} \right) \left(2.00 \times 10^{-9} \text{ C}\right) \left[\frac{-2.00 \times 10^{-6} \text{ C}}{(0.0400 \text{ m} - 0.0100 \text{ m})^2} - \frac{1.00 \times 10^{-6} \text{ C}}{(0.0500 \text{ m} - 0.0400 \text{ m})^2} - \frac{3.00 \times 10^{-6} \text{ C}}{(0.0800 \text{ m} - 0.0400 \text{ m})^2} - \frac{-1.00 \times 10^{-6} \text{ C}}{(0.140 \text{ m} + 0.0400 \text{ m})^2} \right] = -0.252 \text{ N (right)},$$

or $F = \boxed{0.252 \text{ N to the left}}$

(b) The only possible location where the total electric field could be zero is between 5.00 and 8.00 cm, since in that range the two closest charges create forces on a test charge in opposite directions. So, that is the only region we will consider. For the total electric field to be zero between 5.00 and 8.00 cm, we know that:

$$E = \frac{kq_1}{r_1^2} + \frac{kq_5}{r_5^2} - \frac{kq_8}{r_8^2} - \frac{kq_{14}}{r_{14}^2} = 0 .$$

Dividing by common factors and ignoring units (but remembering that x has units of cm), we can get a simplified expression:

$$y = \frac{-2}{(x-1)^2} + \frac{1}{(x-5)^2} - \frac{3}{(x-8)^2} - \frac{-1}{(x-14)^2} .$$

We can then graph this function, using a graphing calculator or graphing program, to determine the values of x that yield $y = 0$.

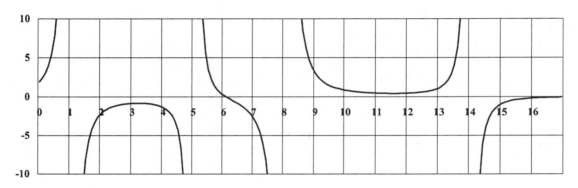

Therefore, the total electric field is zero at $\boxed{x = 6.07 \text{ cm}}$.

17.55 (a) To determine the electric field at the center, we first must determine the distance from each of the charges to the center of the triangle. Since the triangle is equilateral, the center of the triangle will be half way across the base and 1/3 of the way up the height. To determine the height, use the Pythagorean theorem, or the height is given by: $h = \sqrt{(25.0 \text{ cm})^2 - (12.5 \text{ cm})^2} = 21.7 \text{ cm}$.

So, the distance from each charge to the center of the triangle is 2/3 of 21.7 cm, or $r = \frac{2}{3}(21.7 \text{ cm}) = 14.4 \text{ cm}$.

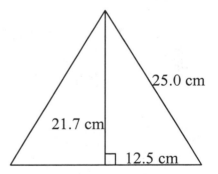

Now that we know the distance from the charges, we can use Equation 17.4 to determine the electric field at the center of the triangle due to each charge: $E = k\dfrac{Q}{r^2}$. Remembering that the electric field is a vector, pointing along the line from the center of the triangle to each charge, in the direction a positive test charge would feel a force, so that:

$$\vec{E}_a = k\frac{q_a}{r^2} = (9.00 \times 10^9 \text{ N} \cdot \text{m}^2/\text{C}^2)\frac{+2.50 \times 10^{-9} \text{ C}}{(0.144 \text{ m})^2} = 1085 \text{ N/C, at a } 90° \text{ angle below the horizontal},$$

$$\vec{E}_b = k\frac{q_b}{r^2} = (9.00 \times 10^9 \text{ N} \cdot \text{m}^2/\text{C}^2)\frac{-8.00 \times 10^{-9} \text{ C}}{(0.144 \text{ m})^2} = 3472 \text{ N/C, at a } 30° \text{ angle below the horizontal, and}$$

$$\vec{E}_c = k\frac{q_c}{r^2} = (9.00 \times 10^9 \text{ N} \cdot \text{m}^2/\text{C}^2)\frac{+1.50 \times 10^{-9} \text{ C}}{(0.144 \text{ m})^2} = 681.0 \text{ N/C, at a } 30° \text{ angle above the horizontal}.$$

Adding the vectors by components gives:

$$E_x = E_a \cos(-90°) + E_b \cos(-30°) + E_c \cos 30°$$

$$E_x = 0 \text{ N/C} + 3472 \text{ N/C}(0.8660) + 681.0 \text{ N/C}(0.8660) = 3597 \text{ N/C}$$

$$E_y = E_a \sin(-90°) + E_b \sin(-30°) + E_c \sin 30°$$

$$E_y = -1085 \text{ N/C} + 3472 \text{ N/C}(-0.5000) + 681.0 \text{ N/C}(0.5000) = -2481 \text{ N/C}$$

so that the electric field is given by:

$$E = \sqrt{E_x^2 + E_y^2} = \sqrt{(3597 \text{ N/C})^2 + (-2481 \text{ N/C})^2} = 4370 \text{ N/C} \text{ and } \theta = \tan^{-1}\frac{E_y}{E_x} = \tan^{-1}\frac{-2481 \text{ N/C}}{3597 \text{ N/C}} = -34.6°,$$

or $\boxed{\vec{E} = 4.37 \times 10^3 \text{ N/C, } 34.6° \text{ below the horizontal}}$.

(b) No, there are no combinations, other than $q_a = q_b = q_c$ that will produce a zero strength electric field at the center of the triangular configuration because of the vector nature of the electric field. Consider the two cases: (1) all charges have the same sign and (2) one charge has a different sign than the other two. For case (1), symmetry dictates that the charges must be all the same magnitude, if a test charge is not to feel a force at the center of the triangle. For case (2), a positive test charge would feel a force toward the negative charge(s) and away from the positive charge(s), therefore there is no combination that would produce a zero strength electric field at the center of the triangle.

17.61

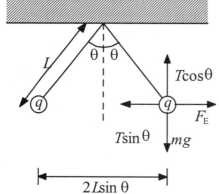

There are three forces acting on each insulating ball, the tension in the string, the force of gravity, and the electric force. The tension acts along the string and upward, gravity acts straight down, and the electric force acts to horizontally outward. Using Newton's Laws, we can get expressions that allow us to solve for the charge on the balls:

$$\text{net } F_x = T \cos \theta - mg = 0 \Rightarrow T = \frac{mg}{\cos \theta}$$

$$\text{net } F_y = F_E - T \sin \theta = 0 \Rightarrow F_E = T \sin \theta = mg \tan \theta$$

Now, using Equation 17.2, we can rewrite the electric force in terms of the charges and the distance separating the charges:

$$F_E = k\frac{q^2}{r^2} = k\frac{q^2}{(2L \sin \theta)^2}$$

now, since we have another expression for the electric force, we can solve for the charge:

$$\frac{kq^2}{(2L \sin \theta)^2} = mg \tan \theta \text{ so that } q = \sqrt{\frac{mg \tan \theta}{k}}(2L \sin \theta)$$

or substituting in the known values gives:

$$q = \sqrt{\frac{(0.500 \times 10^{-3} \text{ kg})(9.80 \text{ m/s}^2)(\tan 10°)}{9.00 \times 10^9 \text{ N} \cdot \text{m}^2/\text{C}^2}}[2(0.200 \text{ m})(\sin 10°)] = 2.15 \times 10^{-8} \text{ C} = \boxed{21.5 \text{ nC}}$$

17.67 (a) To determine the acceleration, use Newton's Laws and Equation 17.2:

$$F = ma = \frac{kq_1 q_2}{r^2} \Rightarrow a = \frac{kq^2}{mr^2} = \frac{(9.00 \times 10^9 \text{ N} \cdot \text{m}^2/\text{C}^2)(1.00 \times 10^{-3} \text{ C})^2}{(0.500 \times 10^{-3} \text{ kg})(0.0100 \text{ m})^2} = \boxed{1.80 \times 10^{11} \text{ m/s}^2}$$

(b) The resulting acceleration is unreasonably large; the raindrops would not stay together.

(c) The assumed charge of 1.00 mC is much too great; typical static electricity is on the order of 1 μC, or less.

ELECTRIC POTENTIAL AND ELECTRIC ENERGY

18

CONCEPTUAL QUESTIONS

18.1 Potential difference is more descriptive because the impact on the charges is really dependent on the difference in potential. Voltage is the common word because the units of potential difference is volts.

18.4 It is the difference in voltage between two points that tells you how a charge behaves while moving between those two points.

18.7 If the electric potential is constant, then the potential difference between two points is zero, thus from Equation 18.5, the electric field is zero.

18.10 The potential due to a uniformly charged sphere is the same as that of a point charge outside the sphere. Inside the sphere, the potential is different from that of a point charge.

18.13 You can think of equipotential lines as contour lines on a map. They represent points where the voltage "height" is the same. Electric field lines show the steepest path down these hills. Thus, equipotential lines show "sameness", while electric fields show "maximum change", resulting in the potentials and electric field lines being perpendicular.

18.16 In a capacitor, you need to store an excess amount of like charges on a plate. Coulomb's law says you must apply more force the closer you pack like charges. Increasing the area allows you to spread charge out more, and hense increases the ability to store charge, i.e. the capacitance. Similarly, since opposite charges attract and the plates of a capacitor hold opposite charges, the closer the plates are the more charge that can be stored, i.e. the larger the capacitance. So, the capacitance should be proportional to area and inversely proportional to separation distance.

18.19 The polar nature of the water means that the water molecules in the air will orient themselves to effectively reduce the separation distance between the plates, making it easier for a spark to cross the gap.

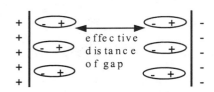

18.22 Comparing Equations 18.13 and 18.14, we find that the total capacitance is greatest when the capacitors are placed in parallel. Since for a fixed potential difference the energy is proportional to the capacitance (from Equation 18.15), the largest energy will be when the configuration has the largest capacitance, or when the capacitors are placed in parallel.

PROBLEMS

18.1 Using Equation 18.3, we can calculate the change in potential energy, given the potential difference and the charge:

$$\Delta PE = q\Delta V = \left(125 \times 10^{-9} \text{ C}\right)(5000 \text{ V}) = \boxed{6.25 \times 10^{-4} \text{ J}}$$

18.7   Recall from Equation 6.9 that power is work divided by time: $P = \dfrac{W}{t}$. Now, using Equation 18.3, $\Delta PE = q\Delta V$

we can express the work (or change in potential energy) in terms of the charge and the voltage of the battery:

$$P = \frac{qV}{t} = \frac{(2.00\ \text{C})(9.00\ \text{V})}{3600\ \text{s}} = \boxed{5.00 \times 10^{-3}\ \text{W}}$$

18.13   To determine the voltage needed to accelerate the negative hydrogen ion, we first need to remember that there will be a change in potential energy because the ion will increase its kinetic energy:

$$\Delta PE = KE_f - KE_i = \frac{1}{2}mv_f^2 - \frac{1}{2}\cancel{mv_i^2} = \frac{1}{2}mv_f^2 .$$

So, now that we know how the potential energy changes, we can use Equation 18.3 to determine the voltage required to accelerate the ion to 1.00% of the speed of light:

$$\Delta V = \frac{\Delta PE}{q} = \frac{mv_f^2}{2q} = \frac{\left(1.67 \times 10^{-27}\ \text{kg}\right)\left[(0.0100)\left(3.00 \times 10^8\ \text{m/s}\right)\right]^2}{2\left(-1.60 \times 10^{-19}\ \text{C}\right)} = -4.70 \times 10^4\ \text{V} .$$

Therefore, a voltage of $\boxed{47.0\ \text{kV}}$ is required to accelerate the ion.

18.19   (a)   Using Equation 18.5, we can determine the electric field strength produced between two parallel plates since we know their separation distance and the potential difference across the plates:

$$E = \frac{V_{AB}}{d} = \frac{5000\ \text{V}}{2.00 \times 10^{-3}\ \text{m}} = \boxed{2.50 \times 10^6\ \text{V/m}} < 3 \times 10^6\ \text{V/m} .$$

$\boxed{\text{No}}$, the field strength is smaller than the breakdown strength for air.

(b)   Using Equation 18.5, we can now solve for the separation distance, given the potential difference and the maximum electric field strength:

$$d = \frac{V_{AB}}{E} = \frac{5000\ \text{V}}{3 \times 10^6\ \text{V/m}} = 1.67 \times 10^{-3}\ \text{m} = \boxed{1.7\ \text{mm}} .$$

So, the plates must not be closer than 1.7 mm to avoid exceeding the breakdown strength of air. Note: the answer is reported only to two digits because the maximum field strength is approximate.

18.25   (a)   Using Equation 18.3, we can get an expression for the change in energy of the electron in terms of the potential difference and its charge: $\Delta KE = q\Delta V$. Also, we know from Equation 18.5, that we can express the potential difference in terms of the electric field strength and the distance traveled, so that:

$$\Delta KE = qV_{AB} = qEd = \left(1.60 \times 10^{-19}\ \text{C}\right)\left(2.00 \times 10^6\ \text{V/m}\right)(0.400\ \text{m})\left(\frac{1\ \text{eV}}{1.60 \times 10^{-19}\ \text{J}}\right)\left(\frac{1\ \text{keV}}{1000\ \text{eV}}\right) = \boxed{800\ \text{keV}} .$$

In other words, the electron would gain 800 keV of energy if accelerated over a distance of 0.400 m.

(b)   Using the same expression in part (a), we can now solve for the distance traveled:

$$d = \frac{\Delta KE}{qE} = \frac{\left(50.0 \times 10^9\ \text{eV}\right)}{\left(1.60 \times 10^{-19}\ \text{C}\right)\left(2.00 \times 10^6\ \text{V/m}\right)}\left(\frac{1.60 \times 10^{-19}\ \text{J}}{1\ \text{eV}}\right) = 2.50 \times 10^4\ \text{m} = \boxed{25.0\ \text{km}} .$$

So, the electron must be accelerated over a distance of 25.0 km to gain 50.0 GeV of energy.

18.31 Given Equation 18.8, $V = \dfrac{kQ}{r}$, we can determine the charge given the potential and the separation distance:

$$Q = \frac{rV}{k} = \frac{(15.0 \text{ m})(500 \text{ V})}{9.00 \times 10^9 \text{ N} \cdot \text{m}^2/\text{C}^2} = \boxed{8.33 \times 10^{-7} \text{ C}} \ .$$

The charge is positive because the potential is positive.

18.37 To draw the equipotential lines, remember that they are always perpendicular to electric field lines. The potential is greatest (most positive) near the positive charge, q_2, and least (most negative) near the negative charge, q_1. In other words, the potential increases as you move out from the charge q_1, and it increases as you move into the charge q_2.

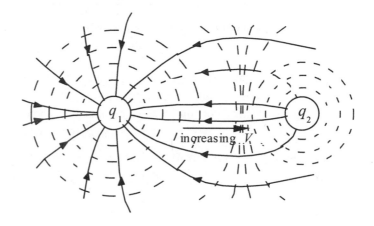

18.43 Using Equation 18.9, we can determine the charge on a capacitor, since we are given its capacitance and its voltage:

$$Q = CV = (180 \times 10^{-6} \text{ F})(120 \text{ V}) = \boxed{2.16 \times 10^{-2} \text{ C}}$$

18.49 Using Equation 18.9, we can determine the voltage that must be applied to a capacitor, given the charge it stores and its capacitance:

$$V = \frac{Q}{C} = \frac{0.160 \times 10^{-3} \text{ C}}{8.00 \times 10^{-9} \text{ F}} = 2.00 \times 10^4 \text{ V} = \boxed{20.0 \text{ kV}}$$

18.55 There are two ways in which you can connect two capacitors: in parallel and in series. When connected in series, the total capacitance is given by Equation 18.13:

$$\frac{1}{C_s} = \frac{1}{C_1} + \frac{1}{C_2} \Rightarrow C_s = \frac{C_1 C_2}{C_1 + C_2} = \frac{(5.00 \ \mu\text{F})(8.00 \ \mu\text{F})}{5.00 \ \mu\text{F} + 8.00 \ \mu\text{F}} = \boxed{3.08 \ \mu\text{F (series)}}$$

and when connected in parallel, the total capacitance is given by Equation 18.14:

$$C_p = C_1 + C_2 = 5.00 \ \mu\text{F} + 8.00 \ \mu\text{F} = \boxed{13.0 \ \mu\text{F (parallel)}}$$

18.61 (a) If the capacitors are connected in series, their total capacitance is given by Equation 18.13:

$$\frac{1}{C_s} = \frac{1}{C_1} + \frac{1}{C_2} \Rightarrow C_s = \frac{C_1 C_2}{C_1 + C_2} = \frac{(2.00 \ \mu F)(7.40 \ \mu F)}{9.40 \ \mu F} = 1.574 \ \mu F \ .$$

Then, since we know the capacitance and the voltage of the battery, we can use Equation 18.9 to determine the charge stored in the capacitors:

$$Q = C_s V = \left(1.574 \times 10^{-6} \ F\right)(9.00 \ V) = \boxed{1.42 \times 10^{-5} \ C} \ .$$

To determine the energy stored in the capacitors, use Equation 18.15:

$$E_{cap} = \frac{C_s V^2}{2} = \frac{\left(1.574 \times 10^{-6} \ F\right)(9.00 \ V)^2}{2} = \boxed{6.38 \times 10^{-5} \ J} \ .$$

Note: by using the form of Equation 18.15 involving capacitance and voltage, we can avoid using one of the parameters that we calculated, minimizing our chances of propagating an error.

(b) If the capacitors are connected in parallel, their total capacitance is given by Equation 18.14:

$$C_p = C_1 + C_2 = 2.00 \ \mu F + 7.40 \ \mu F = 9.40 \ \mu F \ .$$

Again, we use Equation 18.9 to determine the charge stored in the capacitors:

$$Q = C_p V = \left(9.40 \times 10^{-6} \ F\right)(9.00 \ V) = \boxed{8.46 \times 10^{-5} \ C} \ .$$

And finally, using Equation 18.15 again, we can determine the energy stored in the capacitors:

$$E_{cap} = \frac{C_p V^2}{2} = \frac{\left(9.40 \times 10^{-6} \ F\right)(9.00 \ V)^2}{2} = \boxed{3.81 \times 10^{-4} \ J}$$

18.67

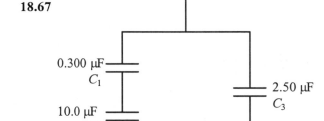

To determine the capacitance, first combine capacitors C_1 and C_2, which are in series (so use Equation 18.13):

$$C_s = \frac{C_1 C_2}{C_1 + C_2} = \frac{(0.30 \ \mu F)(10 \ \mu F)}{(0.30 \ \mu F + 10 \ \mu F)} = 0.2913 \ \mu F \ .$$

Then combine C_s and C_3, which are in parallel (using Equation 18.14):

$$C = C_s + C_3 = 0.2913 \ \mu F + 2.5 \ \mu F = 2.791 \ \mu F \ .$$

The voltage across the whole assembly is equal to the voltage across C_3, or the voltage across C_1 and C_2, so we'll use C_3 to determine the voltage since its easier. Using Equation 18.15, we can solve for the voltage given the energy stored in capacitor C_3:

$$V = \left(\frac{2E_3}{C_3}\right)^{1/2} = \left[\frac{2\left(1.80 \times 10^{-4} \ J\right)}{2.50 \times 10^{-6} \ F}\right]^{1/2} = 12.0 \ V \ .$$

Finally, since we know the voltage drop, we can determine the total energy stored in the capacitors using Equation 18.15, where the capacitance is the total capacitance, C, found above:

$$E = \frac{C V^2}{2} = \frac{\left(2.791 \times 10^{-6} \ F\right)(12.0 \ V)^2}{2} = \boxed{2.01 \times 10^{-4} \ J} \ .$$

18.73 (a)   Recall from Equation 6.9, that the power is the work divided by the time, so the average power is:

$$P = \frac{W}{t} = \frac{400 \text{ J}}{10.0 \times 10^{-3} \text{ s}} = \boxed{4.00 \times 10^4 \text{ W}}.$$

(b)   A defibrillator does not cause serious burns because the skin conducts electricity well at high voltages, like those used in defibrillators. The jell used aids in the transfer of energy to the body, and the skin doesn't absorb the energy, but rather, lets it pass through to the heart.

18.79   In order for the drop to be suspended, the weight force must equal the Coulomb force (Equation 17.3), so that

$$mg = qE \text{ , or } E = \frac{mg}{q} = \frac{\rho V g}{q}.$$

Now, to determine the voltage between the metal plates, we use Equation 18.5: $V = Ed$. Combining all the known information gives:

$$V = \frac{\rho V g d}{q} = \left(920 \text{ kg/m}^3\right)\left[\frac{4}{3}\pi\left(1.00 \times 10^{-6} \text{ m}\right)^3\right]\frac{\left(9.80 \text{ m/s}^2\right)\left(0.0250 \text{ m}\right)}{1.60 \times 10^{-19} \text{ C}} = \boxed{5.90 \times 10^3 \text{ V}}$$

19 ELECTRIC CURRENT, RESISTANCE, AND OHM'S LAW

CONCEPTUAL QUESTIONS

19.1 Yes, current is the *motion* of charge. For example, in a closed loop of current carrying wire, the negative charges are actually moving around the loop, while the positive charges remain fixed, but there is a neutral charge distribution because the negative charge distribution is the same at all times, i.e. evenly distributed around the loop.

19.4 Two paths are needed because you need a closed loop. If just one path were present, there would be a buildup of charge until it was at the same potential as the source, then no more current would flow. The second path allows the current to continue flowing because it allows a path for current to "drain".

19.7 No, every electron that enters the resistor "pushes" an electron out the other side. *IR* is the amount of "push" needed, but the resistor does not absorb charge or change the current.

19.10 Yes, because of Equation 19.6. Changing the path changes the length, *L*, and the cross-sectional area, *A*, thus changing the resistance.

19.13 Over time, the cross-sectional area of the filament decreases, and eventually it is so thin that it breaks. According to Equation 19.6, as the cross-sectional area decreases, the resistance increases. Then, since the voltage is kept constant, Equation 19.9b tells us that the power will decrease as the resistance increases. If the power decreases, the light bulb grows dim, which happens late in its life.

19.16 For AC electricity, a full cycle goes through zero twice. Therefore, for 60 Hz AC electricity, that is for electricity that goes through 60 cycles per second, the voltage, current and power will go through zero twice as often, or 120 times per second.

19.19 The two types of electric hazards are thermal (excessive power) and shock (current through a person).

19.22 This provides a lot of different paths for the current to flow through.

19.25 Wet skin allows salts to dissolve and that makes a good ionic conductor. Blood and other bodily fluids have mobile ions (from salts) that can carry current.

19.28 Even though both the coulomb force and concentration gradient would imply that the Na^+ ions would move from the outside to the inside, the membrane, which is impermeable to the Na^+ ions prevents the migration.

PROBLEMS

19.1 Using Equation 19.1, we can calculate the current given the charge and the time, remembering that 1 A = 1 C/s:

$$I = \frac{\Delta Q}{\Delta t} = \frac{4.00 \text{ C}}{4.00 \text{ h}}\left(\frac{1 \text{ h}}{3600 \text{ s}}\right) = 2.778 \times 10^{-4} \text{ A} = \boxed{0.278 \text{ mA}}$$

19.7 (a) Using Equation 19.4, we can calculate the resistance of the path given the current and the potential:

$$I = \frac{V}{R}, \text{ so that } R = \frac{V}{I} = \frac{10,000 \text{ V}}{6.00 \text{ A}} = 1.667 \times 10^3 \ \Omega = \boxed{1.67 \text{ k}\Omega}$$

(b) If a 50 times larger resistance existed, keeping the current about the same, the power would be increased by a factor of about 50 (see Equation 19.9c), causing much more energy to be transferred to the skin, which could cause serious burns. The gel used reduces the resistance, and therefore reduces the power transferred to the skin.

19.13 (a) Since we know that a He^{++} ion has a charge of twice the basic unit of charge, we can convert the current, which has units of C/s, into the number of He^{++} ions per second:

$$\left(0.250 \times 10^{-3} \text{ C/s}\right)\frac{1 \text{ He}^{++}}{2\left(1.60 \times 10^{-19} \text{ C}\right)} = \boxed{7.81 \times 10^{14} \text{ He}^{++}/\text{s}}$$

(b) Using Equation 19.1, we can determine the time it takes to transfer 1.00 C of charge, since we know the current:

$$I = \frac{\Delta Q}{\Delta t}, \text{ so that } \Delta t = \frac{\Delta Q}{I} = \frac{1.00 \text{ C}}{0.250 \times 10^{-3} \text{ A}} = \boxed{4.00 \times 10^3 \text{ s}}$$

(c) Using our result from part (a), we can determine the time it takes to transfer 1.00 mol of He^{++} ions by converting units:

$$\left(1.00 \text{ mol He}^{++}\right)\left(\frac{6.02 \times 10^{23} \text{ ions}}{\text{mol}}\right)\left(\frac{1 \text{ s}}{7.81 \times 10^{14} \text{ He}^{++} \text{ ions}}\right) = \boxed{7.71 \times 10^8 \text{ s}}$$

19.19 Using Equation 19.4, given the voltage and the current, we can determine the resistance:

$$I = \frac{V}{R}, \text{ so that } R = \frac{V}{I} = \frac{1.35 \text{ V}}{0.200 \times 10^{-3} \text{ A}} = 6.75 \times 10^3 \ \Omega = \boxed{675 \text{ k}\Omega}$$

19.25 We know we want to use Equation 19.6, so we need to determine the radius because the cross-sectional area is $A = \pi r^2$. Since we know the diameter of the wire is 8.252 mm, we can determine the radius of the wire:

$$r = \frac{d}{2} = \frac{8.252 \times 10^{-3} \text{ m}}{2} = 4.126 \times 10^{-3} \text{ m}.$$

We also know, from Table 19.1 that the resistivity of copper is $1.72 \times 10^{-8} \ \Omega \cdot \text{m}$. So, substituting into Equation 19.6 gives a resistance of:

$$R = \frac{\rho L}{A} = \frac{\left(1.72 \times 10^{-8} \ \Omega \cdot \text{m}\right)\left(1.00 \times 10^3 \text{ m}\right)}{\pi\left(4.126 \times 10^{-3} \text{ m}\right)^2} = \boxed{0.322 \ \Omega}.$$

19.31 We can use Equation 19.8 to determine the temperature coefficient of resistivity of the material. Then, by examining Table 19.2, we can determine the type of material used to make the resistor. Since,

$$R = R_0\left(1 + \alpha \Delta T\right) = 1.400 R_0,$$

for a temperature change of $80.0°C$, we can determine α:

$$\alpha \Delta T = 1.400 - 1 \Rightarrow \alpha = \frac{0.400}{\Delta T} = \frac{0.400}{80.0°C} = 5.00 \times 10^{-3}/°C.$$

So, based on the values of α in Table 19.2, the resistor is made of $\boxed{\text{iron}}$.

19.37 (a) We are given the information that $R = 0.820R_0$, when $\Delta T = T - 37.0°C$. Using Equation 19.8, we can determine the value of T. We know then that:

$$R = R_0\left[1 + \alpha\left(T - 37.0°C\right)\right] = 0.820R_0\,,$$

so since $\alpha = -0.0600/°C$, we can solve for the temperature. Dividing by R_0 gives:

$$1 + \alpha\left(T - 37.0°C\right) = 0.820\,,$$

so that $0.180 = -\alpha\left(T - 37.0°C\right)$, or $\left(T - 37.0°C\right) = \dfrac{-0.180}{\alpha} = \dfrac{-0.180}{-0.0600/°C} = 3.00°C$. Finally,

$$T = 37.0°C + 3.00°C = \boxed{40.0°C}\,.$$

(b) If α is negative at low temperatures, then the term $\left[1 + \alpha\left(T - 37.0°C\right)\right]$ in Equation 19.8 can become negative, which implies that the resistance has the opposite sign of the initial resistance, or it has become negative. Since it is not possible to have a negative resistance, the temperature coefficient of resistivity cannot remain negative to low temperatures. In this example, $\boxed{\alpha > \dfrac{1}{37.0°C - T}}$.

19.43 Starting with the equation $P = IV$, we can get an expression for a watt in terms of current and voltage:

$$[P] = W,\ [IV] = A \cdot V = (C/s)(J/C) = J/s = W\,,$$

so that a watt is equal to an ampere-volt.

19.49 Using Equation 19.13b, we can determine the rms voltage, given the peak voltage:

$$V_{rms} = \frac{V_0}{\sqrt{2}} = \frac{679\text{ V}}{\sqrt{2}} = \boxed{480\text{ V}}$$

19.55 (a) Using Equation 19.9a, we can determine the power given the current and the voltage:

$$P = IV = \left(2.00 \times 10^{-3}\text{ A}\right)\left(15.0 \times 10^{3}\text{ V}\right) = \boxed{30.0\text{ W}}$$

(b) Now, using Equation 19.4, we can solve for the resistance, without using the result from part (a):

$$I = \frac{V}{R} \Rightarrow R = \frac{V}{I} = \frac{15.0 \times 10^{3}\text{ V}}{2.00 \times 10^{-3}\text{ A}} = 7.50 \times 10^{6}\ \Omega = \boxed{7.50\text{ M}\Omega}\,.$$

Note, this assume the cauterizer obeys Ohm's law, which will be true for ohmic materials like good conductors.

19.61 Using Equation 19.16b, we can calculate the average power given the rms values for the current and voltage:

$$P_{ave} = I_{rms}V_{rms} = (10.0\text{ A})(120\text{ V}) = 1.20\text{ kW}\,.$$

Next, since the peak power is the peak current times the peak voltage:

$$P_0 = I_0V_0 = 2\left(\frac{1}{2}I_0V_0\right) = 2P_{ave} = \boxed{2.40\text{ kW}}$$

19.67 (a)   From Equation 19.10, we know how the voltage changes with time for an alternating current (AC):
$$V = V_0 \sin 2\pi ft \ .$$

So, if we want the voltage to be equal to $V_0 / 2$, we know that: $\dfrac{V_0}{2} = V_0 \sin 2\pi ft$, so that:

$$\sin 2\pi ft = \frac{1}{2}, \text{ or } t = \frac{\sin^{-1}(0.5)}{2\pi f} \ .$$

Since we have a frequency of 60 Hz, we can solve for the time that this first occurs (remembering to have your calculator in radians!):

$$t = \frac{0.5236 \text{ rad}}{2\pi(60 \text{ Hz})} = 1.39 \times 10^{-3} \text{ s} = \boxed{1.39 \text{ ms}}$$

(b)   Similarly, for $V = V_0 / 2$: $V = V_0 \sin 2\pi ft = V_0$, so that:

$$t = \frac{\sin^{-1} 1}{2\pi f} = \frac{\pi/2 \text{ rad}}{2\pi(60 \text{ Hz})} = 4.17 \times 10^{-3} \text{ s} = \boxed{4.17 \text{ ms}}$$

(c)   Finally, for $V = 0$: $V = V_0 \sin 2\pi ft = 0$, so that $2\pi ft = 0$, π, 2π, ..., or for the first time after $t = 0$:

$$2\pi ft = \pi, \text{ or } t = \frac{1}{2(60 \text{ Hz})} = 8.33 \times 10^{-3} \text{ s} = \boxed{8.33 \text{ ms}} \ .$$

19.73 From Table 19.3, we know that the threshold of sensation is $I = 1.00 \text{ mA}$. The minimum resistance for the shock to not be felt will occur when I is equal to this value. So, using Equation 19.4, $I = \dfrac{V}{R}$, we can determine the minimum resistance for 120 V AC current:

$$R = \frac{V}{I} = \frac{120 \text{ V}}{1.00 \times 10^{-3} \text{ A}} = \boxed{1.20 \times 10^5 \ \Omega} \ .$$

19.79 (a)   From Equation 19.9a, we can determine the power generated by the vaporizer:
$$P = IV = (3.50 \text{ A})(120 \text{ V}) = 420 \text{ J/s} = 0.420 \text{ kJ/s}$$

and since the vaporizer has an efficiency of 95.0%, the heat that is capable of vaporizing the water is:
$$Q = (0.950)Pt \ ,$$

from Equation 6.9. This heat vaporizes the water according to Equation 13.3, where $L_v = 2256 \text{ kJ/kg}$, from Table 13.2, so that: $(0.950)Pt = mL_v$, or

$$m = \frac{(0.950)Pt}{L_v} = \frac{(0.950)(0.420 \text{ kJ/s})(60.0 \text{ s})}{2256 \text{ kJ/kg}} = 0.0106 \text{ kg} \Rightarrow \boxed{10.6 \text{ g/min}}$$

(b)   If the vaporizer is to run for 8.00 hours, we need to calculate the mass of the water by converting units:

$$m_{\text{required}} = (10.6 \text{ g/min})(8.00 \text{ h})\left(\frac{60 \text{ min}}{1 \text{ h}}\right) = 5.09 \times 10^3 \text{ g} = \boxed{5.09 \text{ kg}} \ .$$

In other words, making use of Table 10.1 to get the density of water, it requires

$$5.09 \text{ kg} \times \frac{m^3}{10^3 \text{ kg}} \times \frac{L}{10^{-3} \text{ m}^3} = \boxed{5.09 \text{ L}} \text{ of water to run overnight.}$$

19.85 (a) Using Equation 19.9a, we can determine the power generated:

$$P = IV = (630 \text{ A})(650 \text{ V}) = 4.10 \times 10^5 \text{ W} = \boxed{410 \text{ kW}}$$

(b) Since the efficiency is 95.0%, the effective power is: $P_{\text{effective}} = (0.950)P = 389.0 \text{ kW}$. Then, using Equation 6.9, we can calculate the work done by the train $W = (P_{\text{eff}})t$. Setting that equal to the change in kinetic energy, Equation 6.2,

$$W = \frac{1}{2}mv^2 - \frac{1}{2}mv_0^2,$$

gives us an expression for the time it takes to reach 20.0 m/s from rest: $W = (P_{\text{eff}})t = \frac{1}{2}mv^2 - \frac{1}{2}mv_0^2$, so that

$$t = \frac{(1/2)mv^2 - (1/2)mv_0^2}{P_{\text{eff}}} = \frac{0.5(5.30 \times 10^4 \text{ kg})(20.0 \text{ m/s})^2}{389 \times 10^3 \text{ W}} = 27.25 \text{ s} = \boxed{27.2 \text{ s}}$$

(c) From Equation 2.9, we recall that $v = v_0 + at$, so that:

$$a = \frac{v - v_0}{t} = \frac{20.0 \text{ m/s}}{27.25 \text{ s}} = \boxed{0.734 \text{ m/s}^2}$$

(d) A typical automobile can go from 0 to 60 mph in 10 seconds, so that its acceleration is:

$$a = \frac{v}{t} = \frac{60 \text{ mi/hr}}{10 \text{ s}} \times \frac{1 \text{ hr}}{3600 \text{ s}} \times \frac{1609 \text{ m}}{\text{mi}} = 2.7 \text{ m/s}^2.$$

Thus, a light-rail train accelerates much slower than a car, but it can reach final speeds substantially faster than a car can sustain. So, typically light-rail tracks are very long and straight, to allow them to reach these faster final speeds without decelerating around sharp turns.

19.91 (a) Using Equation 19.8 and setting the resistance equal to twice the initial resistance, we can solve for the final temperature:

$$R = R_0(1 + \alpha \Delta T) = 2R_0 \Rightarrow \alpha \Delta T = \alpha(T - T_0) = 1,$$

where $T_0 = 20°\text{C}$. So the final temperature will be:

$$T - T_0 = \frac{1}{\alpha} = \frac{1}{0.002 \times 10^{-3}/°\text{C}} = 5 \times 10^5 \text{ °C, so that } T = \boxed{5 \times 10^5 \text{ °C}}.$$

(b) Again, using Equation 19.8, we can solve for the final temperature when the resistance is half the initial resistance:

$$R = R_0(1 + \alpha \Delta T) = \frac{R_0}{2} \Rightarrow \alpha(T - T_0) = -\frac{1}{2},$$

so the final temperature will be:

$$T - T_0 = \frac{-0.5}{0.002 \times 10^{-3}/°\text{C}} = -2.50 \times 10^5 \text{ °C, or } T = -2.5 \times 10^5 \text{ °C} = \boxed{-3 \times 10^5 \text{ °C}}.$$

(c) In part (a), the temperature is above the melting point of any metal. In part (b) the temperature is far below 0 K, which is impossible.

(d) The assumption that the resistivity for Constantin will remain constant over the derived temperature ranges in parts (a) and (b) above is wrong. For large temperature changes, α may vary or a non-linear equation may be needed to find ρ.

CIRCUITS AND DC INSTRUMENTS 20

CONCEPTUAL QUESTIONS

20.1 When the switch is open, there is no current flowing in the wire because there is no closed loop. When the switch is closed, there is a current flowing given by: $I = \dfrac{E}{r+R}$ because of a nearly zero resistance in the switch.

20.4 The power dissipated is small in the closed switch because the resistance is nearly zero, and the power is given by: $P_{switch} = I^2 R_{switch} \quad 0$.

20.7 Yes, the main draw on the power is the headlights themselves and the battery's internal resistance, not the wires connecting them. Replacing the wires by superconductors would reduce the resistance of the wires to zero, but their resistance is already negligible anyway.

20.10 Connect the resistors in parallel. We know from Equation 20.2 that resistors in parallel produce a smaller total resistance than any individual resistance in the combination.

20.13 An emf is a special type of potential difference, therefore every emf is a potential difference, but not every potential difference is an emf.

20.16 The 850 rated battery. The current is given by: $I = \dfrac{V}{r+R}$, so the larger the current the smaller the internal resistance.

20.19 No, Kirchoff's first rule says that the sum of all currents entering a junction must equal the sum of all currents leaving the junction, so if all the currents in Figure 20.41 were entering, then they must all equal zero, since there would be no current leaving the junction.

20.22 Starting at point a and working counterclockwise gives:
$$+I_1 r_1 - E_1 + I_1 R_4 + E_4 + I_2 r_4 + I_2 r_3 - E_3 + I_2 R_3 + I_1 R_1 = 0$$

20.25 The ammeter has a small resistance, so this circuit produces a large current, which will (1) destroy the induction coil in the ammeter and (2) drain the battery.

20.28 (a) Place the ammeter between (ab) or (hg).
(b) Place the ammeter between (bi) or (jg).
(c) Place the ammeter between (bc), (de), or (fg).
(d) Place the ammeter between (bc), (de), or (fg).

20.31 $[\tau] = [R][C] = \Omega \cdot F = \left(\dfrac{V}{A}\right) \cdot \left(\dfrac{C}{V}\right) = \dfrac{C}{A} = \dfrac{C}{(C/s)} = s$

20.34

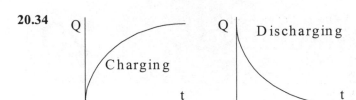

20.37

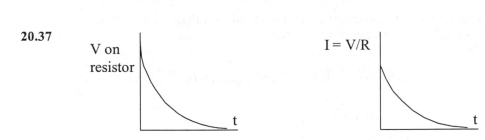

20.40 The resistance of the bleeder must be large compared to the effective resistance of the rest of the circuit or else it will draw too much power from the circuit while it is running This larger resistance will mean the time constant will be large, which is why it is called a "bleeder resistor", since it slowly drops the current from the capacitor. . Also, when the apparatus is shut off, the large resistance of the bleeder will prevent the large capacitor from discharging too quickly preventing a large thunder-like noise, which would be disconcerting.

PROBLEMS

20.1 (a) From Equation 20.1, we know that resistors in series add:

$$R_s = R_1 + R_2 + \ldots + R_{10} = (275\ \Omega)(10) = \boxed{2.75\ \text{k}\Omega}$$

(b) From Equation 20.2, we know that resistors in series add like:

$$\frac{1}{R_p} = \frac{1}{R_1} + \frac{1}{R_2} + \ldots + \frac{1}{R_{10}} = (10)\left(\frac{1}{275\ \Omega}\right) = 3.64 \times 10^{-2}/\Omega,$$

so that

$$R_p = \left(\frac{1}{3.64 \times 10^{-2}}\right)\Omega = \boxed{27.5\ \Omega}$$

20.7 ***Step 1:*** The circuit diagram is drawn in Figure 20.6.

Step 2: Find I_3.

Step 3: Resistors R_2 and R_3 are in parallel. Then, resistor R_1 is in series with the combination of R_2 and R_3.

Step 4: (a) Looking at the point where the wire comes into the parallel combination of R_2 and R_3, we see that the current coming in, I, is equal to the current going out, I_2 and I_3, so that

$$I = I_2 + I_3, \text{ or } I_3 = I - I_2 = 2.35\ \text{A} - 1.61\ \text{A} = \boxed{0.74\ \text{A}}.$$

(b) Using Ohm's Law for R_3, and the voltage for the combination of R_2 and R_3 found in Example 20.3, we can determine the current:

$$I_3 = \frac{V_p}{R_3} = \frac{9.65\ \text{V}}{13.0\ \Omega} = \boxed{0.742\ \text{A}}.$$

Step 5: The result is reasonable because it is smaller than the incoming current, I, and both methods produce the same answer.

20.13 (a) To determine the number simply divide the 9 V by the emf of each cell:
$$9 \text{ V} \div 1.54 \text{ V} = 5.84 \Rightarrow \boxed{6}$$

(b) If six dry cells are put in series, the actual emf is:
$$1.54 \text{ V} \times 6 = \boxed{9.24 \text{ V}}$$

(c) Internal resistance will decrease the terminal voltage because there will be voltage drops across the internal resistance that will not be useful in the operation of the 9 V battery.

20.19 From Example 20.4, we know that $V = 10.0$ V, and $I = 20.0$ A. Using Equation 19.9b to determine the power gives:
$$P = \frac{V^2}{R_L} = \frac{(10.0 \text{ V})^2}{0.500 \text{ }\Omega} = \boxed{200 \text{ W}}.$$
Using Equation 19.9a to determine the power gives:
$$P = IV = (20.0 \text{ A})(10.0 \text{ V}) = \boxed{200 \text{ W}}.$$

20.25 Using the loop rule for loop abcdefgha in Figure 20.22 gives:
$$\boxed{-I_2 R_2 + \mathcal{E}_1 - I_2 r_1 + I_3 R_3 + I_3 r_2 - \mathcal{E}_2 = 0}.$$

20.31 Using the loop rule for loop akledcba in Figure 20.46 gives:
$$\boxed{\mathcal{E}_2 - I_2 r_2 - I_2 R_2 + I_1 R_5 + I_1 r_1 - \mathcal{E}_1 + I_1 R_1 = 0}$$

20.37 We are given $r = 25.0$ Ω, $V = 0.200$ V, and $I = 50.0$ μA. Since the resistors are in series, the total resistance for the voltmeter is found by using Equation 20.1. So using Ohm's Law (Equation 19.4) we can find the resistance, R:
$$R_{tot} = R + r = \frac{V}{I},$$
so that:
$$R = \frac{V}{I} - r = \frac{0.100 \text{ V}}{50.0 \times 10^{-6} \text{ A}} - 25.0 \text{ }\Omega = 1975 \text{ }\Omega = \boxed{1.98 \text{ k}\Omega}$$

20.43 (a)

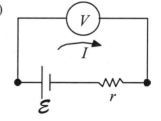

Going counterclockwise around the loop using the loop rule gives:
$$-\mathcal{E} + Ir + IR = 0,$$
or
$$I = \frac{\mathcal{E}}{R + r} = \frac{1.585 \text{ V}}{(1.00 \times 10^3 \text{ }\Omega) + 0.100 \text{ }\Omega}.$$
$$= 1.5848 \times 10^{-3} \text{ A} = \boxed{1.58 \times 10^{-3} \text{ A}}$$

(b) The terminal voltage is given by Equation 20.3:
$$V = \mathcal{E} - Ir = 1.585 \text{ V} - (1.5848 \times 10^{-3} \text{ A})(0.100 \text{ }\Omega) = \boxed{1.5848 \text{ V}}$$
Note: the answer is reported to 5 significant figures to see the difference.

(c) To calculate their ratio, divide the terminal voltage by the emf:
$$\frac{V}{\mathcal{E}} = \frac{1.5848 \text{ V}}{1.585 \text{ V}} = \boxed{0.99990}$$

20.49 Using the method discussed on pages 521-522, we know that: $\mathsf{E}_x = IR_x$ and $\mathsf{E}_s = IR_s$, so that:

$$\frac{\mathsf{E}_x}{\mathsf{E}_s} = \frac{IR_x}{IR_s} = \frac{R_x}{R_s},$$

or

$$\mathsf{E}_x = \mathsf{E}_s\left(\frac{R_x}{R_s}\right) = (1.600\text{ V})\left(\frac{1.200\ \Omega}{1.247\ \Omega}\right) = \boxed{1.540\text{ V}}.$$

20.55 From Equation 20.5, $\tau = RC$, we know that:

$$R_1 = \frac{\tau_1}{C} = \frac{2.00\text{ s}}{0.500 \times 10^{-6}\text{ F}} = 4.00 \times 10^6\ \Omega \text{ and}$$

$$R_2 = \frac{\tau_2}{C} = \frac{15.0\text{ s}}{0.500 \times 10^{-6}\text{ F}} = 3.00 \times 10^7\ \Omega.$$

Therefore, the range for R is: $\boxed{4.00 \times 10^6\ \Omega - 3.00 \times 10^7\ \Omega}$.

20.61 (a) Using Equation 20.5, $\tau = RC$, we can calculate the resistance:

$$R = \frac{\tau}{C} = \frac{10.0 \times 10^{-3}\text{ s}}{8.00 \times 10^{-6}\text{ F}} = 1.25 \times 10^3\ \Omega = \boxed{1.25\text{ k}\Omega}$$

(b) Using Equation 20.6, $V = V_0 e^{-t/RC}$, we can calculate the time it takes for the voltage to drop from 12.0 kV to 600 V:

$$t = -RC\ln\left(\frac{V}{V_0}\right) = -(1.25 \times 10^3\ \Omega)(8.00 \times 10^{-6}\text{ F})\ln\left(\frac{600\text{ V}}{12.0 \times 10^3\text{ V}}\right) = 2.996 \times 10^{-2}\text{ s} = \boxed{30.0\text{ ms}}$$

20.67 From Equation 2.3, we know that $v = \dfrac{x}{t}$, or $t = \dfrac{x}{v}$, and from Equation 20.5, we can write the time constant as $\tau = RC$, so setting these two times equal gives an expression from which we can solve for the required resistance:

$$\frac{x}{v} = RC, \text{ or } R = \frac{x}{vC} = \frac{1.00 \times 10^{-3}\text{ m}}{(500\text{ m/s})(600 \times 10^{-6}\text{ F})} = \boxed{3.33 \times 10^{-3}\ \Omega}$$

20.73 (a) Using Equation 19.9c, $P = I^2 R$, we have $I = \sqrt{\dfrac{P}{R}} = \sqrt{\dfrac{1.00\text{ W}}{15.0\ \Omega}} = 0.258\text{ A}$. So, using Ohm's Law and Equation 20.3, we have $V = \mathsf{E} - Ir = IR$, or:

$$r = \frac{\mathsf{E}}{I} - R = \frac{1.54\text{ V}}{0.258\text{ A}} - 15.0\ \Omega = \boxed{-9.04\ \Omega}$$

(b) You can't have negative resistance!

(c) The voltage should be less than the emf of the battery; otherwise the internal resistance comes out negative. Therefore, the power delivered is too large for the given resistance, leading to a current that is too large.

MAGNETISM ㉑

CONCEPTUAL QUESTIONS

21.1 The material at d_1 emerged and was molten more recently than the material at d_2, which was more recent than d_3, due to the motion of the plates. Assuming the material is magnetized while cooling and it then retains that magnetization after it has cooled and "set", then the earth's magnetic field has

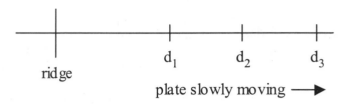

weakened and then reversed periodically over time. By measuring the distance between different magnetizations and knowing the rate at which the plate moves, one can calculate the time over which the local magnetic field weakened and reversed.

21.4 Yes, because electric field strengths decrease rapidly with distance from equal and opposite charges. Yes, this is consistent with how bar magnets behave.

21.7 The magnetic field points to the north outside the sphere of the earth. Therefore, a positively charged particle (i.e. a proton) will have its velocity toward the center of the earth along the equator, and feel a force caused by a magnetic field that points north, so that the right hand rule says that the proton will feel a force to the east and thus be deflected to the east. Similarly, an electron will be deflected to the west. The neutron will not be deflected by the magnetic field because it is uncharged.

21.10 Using the right hand rule: (a) is positive, (b) is neutral, and (c) is negative.

21.13 The greater the mass, the larger the inertia, so the straighter the path. Thus, the particle with the larger mass is particle (a).

21.16 Since $F = ILB$, we need a large ionic current to create a large force to push the fluid. Ocean water is salt water, so it has more ions, making it a better conductor than fresh water. Also, superconducting magnets would be desirable because they have larger magnetic fields, so the force would be larger.

21.19 A DC current would produce a constant magnetic force that would pull the needle. An AC current would cause the needle to oscillate at 120 Hz, which would produce no noticeable effect on the needle.

21.22 No, it is not possible for the middle wire to be repelled by both outside wires or attracted by both outside wires. Equation 21.10 says that "like" currents attract, so if the center wire has a current in the same direction as I_1, it will be attracted to wire I_1 but repelled from wire I_3. Similarly, if the center wire has a current in the same direction as I_3, it will be attracted to wire I_3 but repelled from wire I_1. Thus, it is always attracted to one wire and repelled by the other.

21.25 If the loops were tilted slightly, the attractive force between the wire segments would be stronger for the closer segments and weaker for the farther segments. This would create a torque that would tilt them even more.

21.28 Moving your head quickly could cause dizziness because the ions in your head are aligned by the magnet and don't move with your skull. The strange taste in your mouth is analogous to sticking a 9 V battery on your tongue, which is cause by a small current formed in your mouth.

PROBLEMS

21.1 Use the right hand rule-1 to solve this problem. Your right thumb is in the direction of the velocity, your fingers point in the direction of the magnetic field, and then your palm points in the direction of the magnetic force.

   (a) Your right thumb should be facing down, your fingers out of the page, and then the palm of your hand points to the left (West) .

   (b) Your right thumb should point up, your fingers should point to the right , and then the palm of your hand points into the page .

   (c) Your right thumb should point to the right, your fingers should point into the page, then the palm of your hand points up (North).

   (d) The velocity and the magnetic field are anti-parallel, so there is no force .

   (e) Your right thumb should point into the page, your fingers should point up, and then the palm of your hand points to the right (East) .

   (f) Your right thumb should point out of the page, your fingers should point to the left, and then the palm of your hand points down (South) .

21.7 Examining Equation 21.1, $F = qvB \sin\theta$, we see that the maximum force occurs when $\sin\theta = 1$, so that:
$$F_{max} = qvB = \left(0.100 \times 10^{-6} \text{ C}\right)\left(5.00 \text{ m/s}\right)\left(1.50 \text{ T}\right) = \boxed{7.50 \times 10^{-7} \text{ N}}.$$

21.13 Using Equation 21.3, $r = \dfrac{mv}{qB}$, we can solve for the magnetic field strength necessary to move the proton in a circle of radius 0.800 m:
$$B = \frac{mv}{qr} = \frac{\left(1.67 \times 10^{-27} \text{ kg}\right)\left(7.50 \times 10^{7} \text{ m/s}\right)}{\left(1.60 \times 10^{-19} \text{ C}\right)\left(0.800 \text{ m}\right)} = \boxed{0.979 \text{ T}}$$

21.19 (a)   Since we know Equation 21.3, $r = \dfrac{mv}{qB}$, and we want the radius of the proton to equal the radius of the

electron in Problem 21.12, we can write the velocity of the proton in terms of the information we know about the electron:

$$v_p = \frac{q_p B r}{m_p} = \frac{\cancel{q_p} B}{m_p}\left(\frac{m_e v_e}{\cancel{q_e} B}\right) = \frac{m_e v_e}{m_p} = \frac{\left(9.11\times10^{-31}\ \text{kg}\right)\left(7.50\times10^6\ \text{m/s}\right)}{1.67\times10^{-27}\ \text{kg}} = \boxed{4.09\times10^3\ \text{m/s}}$$

(b)   Now, using Equation 21.3, we can solve for the radius of the proton if the velocity equals the velocity of the electron:

$$r_p = \frac{m v_e}{qB} = \frac{\left(1.67\times10^{-27}\ \text{kg}\right)\left(7.50\times10^6\ \text{m/s}\right)}{\left(1.60\times10^{-19}\ \text{C}\right)\left(1.00\times10^{-5}\ \text{T}\right)} = \boxed{7.83\times10^3\ \text{m}}$$

(c)   First, we need to determine the speed of the proton if the kinetic energies were the same: $\dfrac{1}{2}m_e v_e^2 = \dfrac{1}{2}m_p v_p^2$, so

that

$$v_p = v_e \sqrt{\frac{m_e}{m_p}} = \left(7.5\times10^6\ \text{m/s}\right)\sqrt{\frac{9.11\times10^{-31}\ \text{kg}}{1.67\times10^{-27}\ \text{kg}}} = 1.752\times10^5\ \text{m/s}.$$

Then, using Equation 21.3, we can determine the radius:

$$r = \frac{mv}{qB} = \frac{\left(1.67\times10^{-27}\ \text{kg}\right)\left(1.752\times10^5\ \text{m/s}\right)}{\left(1.60\times10^{-19}\ \text{C}\right)\left(1.00\times10^{-5}\ \text{T}\right)} = \boxed{1.83\times10^2\ \text{m}}.$$

(d)   First, we need to determine the speed of the proton if the momentums are the same: $m_e v_e = m_p v_p$, so that

$$v_p = v_e\left(\frac{m_e}{m_p}\right) = \left(7.50\times10^6\ \text{m/s}\right)\left(\frac{9.11\times10^{-31}\ \text{kg}}{1.67\times10^{-27}\ \text{kg}}\right) = 4.091\times10^3\ \text{m/s}.$$

Then, using Equation 21.3, we can determine the radius:

$$r = \frac{mv}{qB} = \frac{\left(1.67\times10^{-27}\ \text{kg}\right)\left(4.091\times10^3\ \text{m/s}\right)}{\left(1.60\times10^{-19}\ \text{C}\right)\left(1.00\times10^{-5}\ \text{T}\right)} = \boxed{4.27\ \text{m}}.$$

21.25   Using Equation 21.4, $\mathsf{E} = B\ell v$, we can determine the average velocity of the fluid. Note that the width is actually the diameter in this case:

$$v = \frac{\mathsf{E}}{B\ell} = \frac{60.0\times10^{-3}\ \text{V}}{\left(0.500\ \text{T}\right)\left(0.0300\ \text{m}\right)} = \boxed{4.00\ \text{m/s}}$$

21.31   Using Equation 21.4, $\mathsf{E} = B\ell v$, where the width is twice the radius, $\ell = 2r$, and using Equation 19.3, $I = nqAv_d$,

we can get an expression for the drift velocity: $v_d = \dfrac{I}{nqA} = \dfrac{I}{nq\pi r^2}$, so substituting into Equation 21.4 gives:

$$\mathsf{E} = B\times 2r\times\frac{I}{nq\pi r^2} = \frac{2IB}{nq\pi r} \propto \frac{1}{r} \propto \frac{1}{d}.$$

So, the Hall voltage is inversely proportional to the diameter of the wire.

21.37 Using Equation 21.5, where ℓ is the diameter of the tube, we can find the force on the water:
$$F = I\ell B \sin\theta = (100 \text{ A})(0.250 \text{ m})(2.00 \text{ T})(1) = \boxed{50.0 \text{ N}}$$

21.43 (a) Using Equation 21.6, we see that the maximum torque occurs when $\sin\phi = 1$, so the maximum torque is:
$$\tau_{max} = NIAB\sin\phi = (150)(50.0 \text{ A})(0.180 \text{ m})^2(1.60 \text{ T})(1) = \boxed{389 \text{ N}\cdot\text{m}}$$

(b) Now, use Equation 21.6, and set $\phi = 20.0°$, so that the torque is:
$$\tau = NIAB\sin\phi = (150)(50.0 \text{ A})(0.180 \text{ m})^2(1.60 \text{ T})\sin 20.0° = \boxed{355 \text{ N}\cdot\text{m}}$$

21.49 (a)

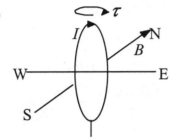

The torque, τ, is CW as seen from directly above, since the loop will rotate CW as seen from directly above. Using Equation 21.6, we find the maximum torque to be:
$$\tau = NIAB = (200)(100 \text{ A})\pi(0.500 \text{ m})^2(3.00\times 10^{-5} \text{ T})$$
$$= \boxed{0.471 \text{ N}\cdot\text{m}}$$

(b) If the loop was connected to a wire, this is an example of a simple motor (see Figure 21.30). When current is passed through the loops, the magnetic field exerts a torque on the loops, which rotates a shaft. Electrical energy is converted to mechanical work in the process.

21.55 Using Equation 21.8, we can calculate the magnetic field strength at the center of a circular loop:
$$B = \frac{\mu_0 I}{2R} = \frac{(4\pi\times 10^{-7} \text{ T}\cdot\text{m/A})(1.05\times 10^4 \text{ A})}{(2)(0.650\times 10^{-15} \text{ m})} = \boxed{1.01\times 10^{13} \text{ T}}.$$

21.61 Using Equation 21.8, $B = \dfrac{\mu_0 I}{2R}$ we can solve for the maximum current in the loop given a magnetic field less than 5.00×10^{-7} T at the center of the wire:
$$I = \frac{2RB}{\mu_0} = \frac{(2)(1.00 \text{ m})(5.00\times 10^{-7} \text{ T})}{4\pi\times 10^{-7} \text{ T}\cdot\text{m/A}} = \boxed{0.796 \text{ A}}$$

21.67 (a) Using Equation 21.10, $\dfrac{F}{\ell} = \dfrac{\mu_0 I_1 I_2}{2\pi r}$, we can calculate the force on the two wires:
$$F = \frac{\ell \mu_0 I^2}{2\pi r} = \frac{(50.0 \text{ m})(4\pi\times 10^{-2} \text{ T}\cdot\text{m/A})(800 \text{ A})^2}{2\pi(0.750 \text{ m})} = \boxed{8.53 \text{ N}}.$$

The force is $\boxed{\text{repulsive}}$ because the currents are in opposite directions.

(b) This force is repulsive and therefore there is never a risk that the two wires will touch and short circuit.

21.73

A
$\odot$ 5.00 A

10.0 A
$\otimes$
B

20.0 A
$\otimes$
C

Opposites repel, likes attract, so we need to consider each wire's relationship with the other two wires. Let f denote force per unit length, then by Equation 21.10: $f = \dfrac{\mu_0 I_1 I_2}{2\pi r}$

$$f_{AB} = \frac{\left(4\pi \times 10^{-7}\ \text{T}\cdot\text{m/A}\right)(5.00\ \text{A})(10.0\ \text{A})}{2\pi(0.100\ \text{m})}$$

$$= 1.00 \times 10^{-4}\ \text{N/m}$$

$$f_{BC} = \frac{\left(4\pi \times 10^{-7}\ \text{T}\cdot\text{m/A}\right)(10.0\ \text{A})(20.0\ \text{A})}{2\pi(0.100\ \text{m})}$$

$$= 4.00 \times 10^{-4}\ \text{N/m} = 4f_{AB}$$

$$f_{AC} = \frac{\left(4\pi \times 10^{-7}\ \text{T}\cdot\text{m/A}\right)(5.00\ \text{A})(20.0\ \text{A})}{2\pi(0.100\ \text{m})} = 2.00 \times 10^{-4}\ \text{N/m} = 2f_{AB}$$

Look at each wire separately

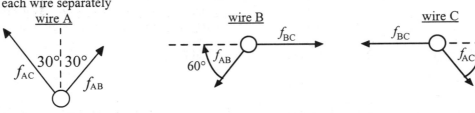

wire A   wire B   wire C

For Wire A:

$$f_{Ax} = f_{AB} \sin 30° - f_{AC} \sin 30° = \left(f_{AB} - 2f_{AB}\right)\sin 30° = -f_{AB}\left(\sin 30°\right) = -0.500 \times 10^{-4}\ \text{N/m}$$

$$f_{Ay} = f_{AB} \cos 30° + f_{AC} \cos 30° = \left(f_{AB} + 2f_{AB}\right)\cos 30° = 3f_{AB}\left(\cos 30°\right) = 2.60 \times 10^{-4}\ \text{N/m}$$

$$F_A = \sqrt{f_{Ax}^2 + f_{Ay}^2} = \boxed{2.65 \times 10^{-4}\ \text{N/m}}$$

$$\theta_A = \tan^{-1}\left(\frac{|f_{Ax}|}{f_{Ay}}\right) = \boxed{10.9°}$$

For Wire B:

$$f_{Bx} = f_{BC} - f_{AB} \cos 60° = 4.00 \times 10^{-4}\ \text{N/m} - \left(1.00 \times 10^{-4}\ \text{N/m}\right)\cos 60° = 3.50 \times 10^{-4}\ \text{N/m}$$

$$f_{By} = -f_{AB} \sin 60° = -\left(1.00 \times 10^{-4}\ \text{N/m}\right)\sin 60° = -0.866 \times 10^{-4}\ \text{N/m}$$

$$F_B = \sqrt{f_{Bx}^2 + f_{By}^2} = \boxed{3.61 \times 10^{-4}\ \text{N/m}}$$

$$\theta_B = \tan^{-1}\left(\frac{|f_{By}|}{f_{Bx}}\right) = \boxed{13.9°}$$

For Wire C:

$$f_{Cx} = f_{AC} \cos 60° - f_{BC} = \left(2.00 \times 10^{-4}\ \text{N/m}\right)\cos 60° - 4.00 \times 10^{-4}\ \text{N/m} = -3.00 \times 10^{-4}\ \text{N/m}$$

$$f_{Cy} = -f_{AC} \sin 60° - f_{BC} = -\left(2.00 \times 10^{-4}\ \text{N/m}\right)\sin 60° - 4.00 \times 10^{-4}\ \text{N/m} = -1.73 \times 10^{-4}\ \text{N/m}$$

$$F_C = \sqrt{f_{Cx}^2 + f_{Cy}^2} = \boxed{3.46 \times 10^{-4}\ \text{N/m}}$$

$$\theta_C = \tan^{-1}\left(\frac{f_{Cy}}{f_{Cx}}\right) = \boxed{30.0°}$$

21.79 (a) Using Equation 10.2, we can calculate the pressure:

$$P = \frac{F}{A} = \frac{F}{\pi r^2} = \frac{50.0 \text{ N}}{\pi(0.125 \text{ m})^2} = \boxed{1.02 \times 10^3 \text{ N/m}^2}$$

(b) $\boxed{\text{No}}$, not a significant fraction of an atmosphere. $\dfrac{P}{P_{\text{atm}}} = \dfrac{1.02 \times 10^3 \text{ N/m}^2}{1.013 \times 10^5 \text{ N/m}^2} = 1.01\%$.

21.85

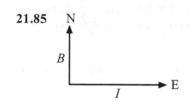

(a) Use the right hand rule-1. Put your right thumb to the East, and your fingers to the North, then your palm points in the direction of the force, or $\boxed{\text{up from ground}}$ (out of page).

(b) Using Equation 21.5, where $\theta = 90°$, so that $F = I\ell B \sin\theta$, or

$$\frac{F}{\ell} = IB \sin\theta = (20.0 \text{ A})(3.00 \times 10^{-5} \text{ T})(1) = \boxed{6.00 \times 10^{-4} \text{ N/m}}$$

(c)

We want the force of the magnetic field to balance the weight force, so $F = mg$.

Now, to calculate the mass, recall Equation 10.1: $\rho = \dfrac{m}{V}$, where the volume is

$V = \pi r^2 L$, so $m = \rho V = \rho \pi r^2 L$ and $F = \rho \pi r^2 L g$, or

$$r = \sqrt{\frac{F/L}{\rho \pi g}} = \sqrt{\frac{6.00 \times 10^{-4} \text{ N/m}}{(8.80 \times 10^3 \text{ kg/m}^3)(\pi)(9.80 \text{ m/s}^2)}} = 4.71 \times 10^{-5} \text{ m} \Rightarrow d = 2r = \boxed{9.41 \times 10^{-5} \text{ m}}$$

(d) From Equation 19.6, we know: $R = \dfrac{\rho L}{A} = \dfrac{\rho L}{\pi r^2}$, where ρ is the resistivity, so that:

$$\frac{R}{L} = \frac{\rho}{\pi r^2} = \frac{1.72 \times 10^{-8} \ \Omega \cdot \text{m}}{\pi(4.71 \times 10^{-5} \text{ m})^2} = 2.47 \ \Omega/\text{m}. \text{ Also, using Equation 19.4, } I = \frac{V}{R}, \text{ we find that}$$

$$\frac{V}{L} = I\frac{R}{L} = (20.0 \text{ A})(2.48 \ \Omega/\text{m}) = \boxed{49.4 \text{ V/m}}.$$

21.91 (a) Using Equation 21.7, $B = \dfrac{\mu_0 I}{2\pi r}$, we can calculate the current required to get the desired magnetic field strength:

$$I = \frac{(2\pi r)B}{\mu_0} = \frac{2\pi(100 \text{ m})(5.00 \times 10^{-5} \text{ T})}{4\pi \times 10^{-7} \text{ T} \cdot \text{m/A}} = 2.50 \times 10^4 \text{ A} = \boxed{25.0 \text{ kA}}$$

(b) This current is unreasonably high. It implies a total power delivery in the line of (from Equation 19.9a):
$$P = IV = (25.0 \times 10^3 \text{ A})(200 \times 10^3 \text{ V}) = 50.0 \times 10^9 \text{ W} = 50.0 \text{ GW}$$
which is much too high for standard transmission lines.

(c) 100 meters is a long distance to obtain the required field strength. Also coaxial cables are used for transmission lines so that there is virtually no field for DC power lines, because of cancellation from opposing currents. The surveyor's concerns are not a problem for his magnetic field measurements.

INDUCTION

22

CONCEPTUAL QUESTIONS

22.1 Multiple loops allow for more area, so by Equation 22.2, there is more induced emf. The iron is ferromagnetic, so it increases the magnetic field, and the induced emf is proportional to the change in flux which is proportional to the magnetic field (see Equations 22.1 and 22.2).

22.4 Yes. The wire has a maximum area when it is in the shape of a circle. Stretching it deforms it to an ellipse of less area. Hence, the current will flow clockwise because of the decreased magnetic flux.

22.7 As the solenoid becomes increasingly electrified, it produces an induced current in the opposite direction in the metal cylinder. The solenoid and the cylinder now act like two oppositely aligned magnets and they repel each other. The cylinder gets hot from Joule heating: $P = I^2 R$, where I is the induced current.

22.10 Assume the particle has a magnetic moment, μ. As the particle passes through the coil, the magnetic flux, Φ, goes from zero to μ and back to zero, this induces a voltage as shown in the diagram.

22.13 As the coils are rotated in a fixed magnetic field, the magnetic flux keeps changing, continually inducing voltage in the coil due to Lenz's law. The coil's equilibrium would be to remain stationary, so work must be done to keep the coil rotating.

22.16 Each coil in the transformer creates a magnetic field, which attracts other current carrying loops. The other current loops are maximally attracted to the coil when the AC current has a peak (either positive or negative), so the oscillations in the other current loops would be twice as often as the AC current's oscillations.

22.19 If the wires are of different length, the rapid increase in current draw when an appliance starts up may not be registered simultaneously. This would cause the GFI to trip.

22.22 Starting with the left side gives: $\dfrac{T \cdot m^2}{A} = \dfrac{[N/A \cdot m] \cdot m^2}{A} = \dfrac{N \cdot m}{A^2}$ and starting with a right side gives:

$$H = \Omega \cdot s = \frac{V}{A} \cdot s = \frac{J/C}{A} \cdot s = \frac{J \cdot s}{A \cdot C} = \frac{(N \cdot m) \cdot s}{A \cdot (A \cdot s)} = \frac{N \cdot m}{A^2}, \text{ so we can then say that: } \frac{T \cdot m^2}{A} = H.$$

22.25 Use a large inductance in series with the computer to filter out the high frequencies. Inductors hate change ($X_L \quad f$, from Equation 22.17), so they will block high frequencies.

22.28 Looking at Equation 22.24, you see that by either increasing the voltage, V, or increasing the power factor, $\cos \phi$, one can increase the average power for an AC circuit. Increasing the voltage, V, however, would increase wasted heat and could burn out the motor. Increasing the power factor, or reducing the phase difference between the current and the voltage, is better because it increases the efficiency of the motor and thus *reduces* wasted heat.

PROBLEMS

22.1   Using Equation 22.1, we can calculate the magnetic flux through coil 2, since the coils are perpendicular:

$$\Phi = BA\cos\theta = BA\cos 90° = \boxed{0}$$

22.7   The units of $\dfrac{\Delta\Phi}{\Delta t}$ will be:

$$\frac{[\Delta\Phi]}{[\Delta t]} = \frac{\text{T}\cdot\text{m}^2}{\text{s}} = (\text{N/A}\cdot\text{m})\left(\frac{\text{m}^2}{\text{s}}\right) = \frac{\text{N}\cdot\text{m}}{\text{A}\cdot\text{s}} = \frac{\text{N}\cdot\text{m}}{\text{C}} = \text{V},$$

so that $\boxed{1\ \text{T}\cdot\text{m}^2/\text{s} = 1\ \text{V}}$.

22.13

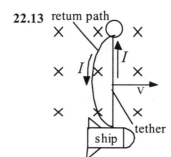

The flux through the loop (into the page) is increasing because the loop is getting larger and enclosing more magnetic field.

Thus, a magnetic field (out of the page) is induced to oppose the change in flux from the original field.

Using RHR-2, put your fingers out of the page within the loop, then your thumb points in the counterclockwise direction around the loop, so the induced magnetic field is produced by the induction of a counterclockwise current in the circuit.

Finally, using RHR-1, putting your right thumb in the direction of the current and your fingers into the page (in the direction of the magnetic field), your palm points to the left, so the magnetic force on the wire is to the left (in the direction opposite to its velocity).

22.19   (a)   The magnetic field is zero and not changing, so there is $\boxed{\text{no current}}$ and therefore $\boxed{\text{no force}}$ on the coil.

(b)   The magnetic field is increasing out of the page, so Lenz's law tells us that the induced magnetic field is into the page. The right hand rule-2 then tells us that this induced magnetic field was created by an induced $\boxed{\text{clockwise current}}$. Finally, the right hand rule-1 applied on the part of the loop exposed to a magnetic field (the right side) tells us that this current creates a $\boxed{\text{force to the left}}$.

(c)   The magnetic field is not changing, so there is $\boxed{\text{no current}}$ and therefore $\boxed{\text{no force}}$ on the coil.

(d)   The magnetic field is decreasing out of the page, so Lenz's law tells us that the induced magnetic field is out of the page. The right hand rule-2 tells us that this induced magnetic field was created by an induced $\boxed{\text{counterclockwise current}}$. Finally, the right hand rule-1 applied on the part of the loop exposed to a magnetic field (the left side) tells us that this current creates a $\boxed{\text{force to the left}}$.

(e)   The magnetic field is zero and not changing, so there is $\boxed{\text{no current}}$ and therefore $\boxed{\text{no force}}$ on the coil.

22.25 From Equation 9.1, and the information given in Problem 22.11:

$$\Delta\theta = \frac{1}{4} \text{ rev} = \frac{1}{4}(2\pi \text{ rad}) \text{ and } \Delta t = 4.17 \times 10^{-3} \text{ s},$$

we can calculate the angular velocity of the coil:

$$\omega = \frac{\Delta\theta}{\Delta t} = \frac{(1/4)(2\pi) \text{ rad}}{4.17 \times 10^{-3} \text{ s}} = 376.7 \text{ rad/s}.$$

Then, using Equation 22.5 and the information given in Problem 22.11:

$$N = 500; \quad A = \pi r^2 = \pi(0.250 \text{ m})^2; \text{ and } B = 0.425 \text{ T},$$

we can calculate the maximum emf generated by the coil:

$$\mathcal{E}_0 = NAB\omega = (500)(\pi)(0.250 \text{ m})^2 (0.425 \text{ T})(376.7 \text{ rad/s}) = 1.57 \times 10^4 \text{ V} = \boxed{15.7 \text{ kV}}$$

22.31 (a) Using Equation 19.4, $I = \dfrac{V}{R}$, we can determine the resistance given the voltage and the current:

$$R = \frac{V}{I} = \frac{120 \text{ V}}{20.0 \text{ A}} = \boxed{6.00 \ \Omega}$$

(b) Again, using Equation 19.4, we can now determine the current given that the net voltage is the difference between the source voltage and the back emf:

$$I = \frac{V}{R} = \frac{120 \text{ V} - 100 \text{ V}}{6.00 \ \Omega} = \boxed{3.33 \text{ A}}$$

22.37 (a) Using Equations 22.6 and 22.7, we can determine the primary current:

$$\frac{I_p}{I_s} = \frac{N_s}{N_p} = \frac{V_s}{V_p},$$

so that:

$$I_p = I_s \left(\frac{V_s}{V_p}\right) = (200 \times 10^{-3} \text{ A})\left(\frac{12.0 \text{ V}}{120 \text{ V}}\right) = 2.00 \times 10^{-2} \text{ A} = \boxed{20.0 \text{ mA}}$$

(b) Using Equation 19.9a, we can calculate the input power using the primary current and voltage:

$$P_{in} = I_p V_p = (2.00 \times 10^{-2} \text{ A})(120 \text{ V}) = \boxed{2.40 \text{ W}}$$

(c) $\boxed{\text{Yes}}$, this amount of power is quite reasonable for a small appliance.

22.43 Using Equation 22.8a, $\mathcal{E}_2 = -M\dfrac{\Delta I_1}{\Delta t}$, where the minus sign is an expression of Lenz's law, we can calculate the mutual inductance between the two coils:

$$M = \mathcal{E}_2 \frac{\Delta t}{\Delta I_1} = (9.00 \text{ V})\frac{(1.00 \times 10^{-3} \text{ s})}{5.00 \text{ A}} = \boxed{1.80 \text{ mH}}$$

22.49 (a) Using Equation 22.9, where the minus sign is an expression of Lenz's law, we can find the induced emf:

$$\mathcal{E} = L\frac{\Delta I}{\Delta t} = (25.0 \text{ H})\frac{(100 \text{ A})}{80.0\times10^{-3} \text{ s}} = 3.125\times10^{4} \text{ V} = \boxed{31.3 \text{ kV}}$$

(b) The energy stored in the inductor can be found using Equation 22.12:

$$E_{\text{ind}} = \frac{1}{2}LI^{2} = \left(\frac{1}{2}\right)(25.0 \text{ H})(100 \text{ A})^{2} = \boxed{1.25\times10^{5} \text{ J}}$$

(c) The question is asking for the power, so using Equation 6.9, we have:

$$P = \frac{\Delta E}{\Delta t} = \frac{1.25\times10^{5} \text{ J}}{80.0\times10^{-3} \text{ s}} = 1.563\times10^{6} \text{ W} = \boxed{1.56 \text{ MW}}$$

(d) $\boxed{\text{No}}$, it is not surprising since this power is very high.

22.55 Using Equation 22.14, $\tau = \dfrac{L}{R}$, we can find the self-inductance for the RL circuit:

$$L = \tau R = (1.00 \text{ s})(500 \text{ }\Omega) = \boxed{500 \text{ H}}$$

22.61 We use Equation 22.13, because the problem says, "after the circuit is completed". Thus, the final current is given by: $I = I_0\left(1 - e^{-t/\tau}\right)$, where $t = 3\tau$ so that:

$$\frac{I}{I_0} = \left(1 - e^{-t/\tau}\right) = 1 - e^{-3} = 0.9502 \ .$$

In other words, the current is $\boxed{95.0\%}$ of the final current three time constants after the circuit is completed.

22.67 Using Equation 22.19, $X_C = \dfrac{1}{2\pi f C}$, we can determine the necessary capacitance:

$$C = \frac{1}{2\pi f X_C} = \frac{1}{2\pi(50.0 \text{ Hz})\left(2.00\times10^{6} \text{ }\Omega\right)} = 1.592\times10^{-9} \text{ F} = \boxed{1.59 \text{ nF}} \ .$$

22.73 (a) Using Equation 22.17, $X_L = 2\pi f L$, we can determine the minimum inductance:

$$L = \frac{X_L}{2\pi f} = \frac{2.00\times10^{3} \text{ }\Omega}{2\pi\left(15.0\times10^{3} \text{ Hz}\right)} = 2.122\times10^{-2} \text{ H} = \boxed{21.2 \text{ mH}} \ .$$

(b) Again using Equation 22.17, this time we can determine the inductive reactance:

$$X_L = 2\pi f L = 2\pi\left(60.0\times10^{3} \text{ Hz}\right)\left(2.122\times10^{-2} \text{ H}\right) = 8.00\times10^{3} \text{ }\Omega = \boxed{8.00 \text{ k}\Omega} \ .$$

22.79 Using Equation 22.22, we can determine the resonant frequency for the circuit:

$$f_0 = \frac{1}{2\pi\sqrt{LC}} = \frac{1}{2\pi\sqrt{\left(0.500\times10^{-3} \text{ H}\right)\left(40.0\times10^{-6} \text{ F}\right)}} = 1.125\times10^{3} \text{ Hz} = \boxed{1.13 \text{ kHz}}$$

22.85 (a) Equation 22.17 gives the inductive reactance:
$$X_L = 2\pi f L = 2\pi (120 \text{ Hz})(100 \times 10^{-6} \text{ H}) = 7.540 \times 10^{-2} \ \Omega$$

Equation 22.19 gives the capacitive reactance:
$$X_C = \frac{1}{2\pi fC} = \frac{1}{2\pi (120 \text{ Hz})(80.0 \times 10^{-6} \text{ F})} = 16.58 \ \Omega$$

Finally, Equation 22.21 gives the impedance of the RLC circuit:
$$Z = \sqrt{R^2 + (X_L - X_C)^2} = \sqrt{(2.50 \ \Omega)^2 + (7.54 \times 10^{-2} \ \Omega - 16.58 \ \Omega)^2} = \boxed{16.7 \ \Omega}$$

(b) Again, Equation 22.17 gives the inductive reactance:
$$X_L = 2\pi (5.00 \times 10^3 \text{ Hz})(100 \times 10^{-6} \text{ H}) = 3.142 \ \Omega$$

Equation 22.19 gives the capacitive reactance:
$$X_C = \frac{1}{2\pi (5.00 \times 10^3 \text{ Hz})(80.0 \times 10^{-6} \text{ F})} = 3.979 \times 10^{-1} \ \Omega$$

And Equation 22.21 gives the impedance:
$$Z = \sqrt{(2.5 \ \Omega)^2 + (3.142 \ \Omega - 3.979 \times 10^{-1} \ \Omega)^2} = \boxed{3.71 \ \Omega}$$

(c) The rms current is found using Equation 22.20. For $f = 120$ Hz,
$$I_{rms} = \frac{V_{rms}}{Z} = \frac{5.60 \text{ V}}{16.69 \ \Omega} = \boxed{0.336 \text{ A}}$$

and for $f = 5.00$ kHz,
$$I_{rms} = \frac{5.60 \text{ V}}{3.712 \ \Omega} = \boxed{1.51 \text{ A}}.$$

(d) The resonant frequency is found using Equation 22.22:
$$f_0 = \frac{1}{2\pi \sqrt{LC}} = \frac{1}{2\pi \sqrt{(1.00 \times 10^{-4} \text{ H})(80.0 \times 10^{-6} \text{ F})}} = 1.779 \times 10^3 \text{ Hz} = \boxed{1.78 \text{ kHz}}$$

(e) At resonance, $X_L = X_R$, so that $Z = R$, and Equation 22.20 reduces to:
$$I_{rms} = \frac{V_{rms}}{R} = \frac{5.60 \text{ V}}{2.50 \ \Omega} = \boxed{2.24 \text{ A}}$$

22.91 Using Equation 20.1, since the resistors are in series, we know the total internal resistance of the batteries is:
$r = 4(0.100 \ \Omega)$. So, Ohm's law becomes:
$$I = \frac{E - V}{r + R},$$

so that,
$$R + r = \frac{E - V}{I}.$$

Therefore, the resistance is:
$$R = \frac{E - V}{I} - r = \frac{6.00 \text{ V} - 4.50 \text{ V}}{3.00 \text{ A}} - 4(0.100 \ \Omega) = \boxed{0.100 \ \Omega}.$$

22.97 (a) From Equation 22.2, we know how the induced emf depends on the magnetic flux:

$$\mathsf{E}_0 = -\frac{N\Delta\Phi}{\Delta t},$$

where the minus sign means that the emf creates a current and magnetic field that opposes the change in flux. From Equation 22.1, we can write the magnetic flux in terms of the magnetic field and the area of the loop:

$$\Phi = BA = \left(\pi r^2\right)B .$$

Since the only thing that varies in the magnetic flux is the magnetic field, we can then say that:

$$\Delta\Phi = \pi r^2 \Delta B .$$

Now, from Equation 21.7, we can express the magnetic field in terms of the current and the distance from the wire:

$$B = \frac{\mu_0 I}{2\pi d},$$

so that the change in magnetic field occurs because of a change in the current, or

$$\Delta B = \frac{\mu_0 \Delta I}{2\pi d} .$$

Finally, substituting back into Equation 22.22 gives:

$$\mathsf{E}_0 = -\frac{N\pi r^2 \mu_0 \Delta I}{2\pi d\,\Delta t} = -\frac{\mu_0 N r^2 \Delta I}{2d\,\Delta t} = -\frac{\left(4\pi\times10^{-7}\ \mathrm{T\cdot m/A}\right)(1)(0.500\ \mathrm{m})^2\left(-2.00\times10^6\ \mathrm{A}\right)}{2(50.0\ \mathrm{m})(25.0\times10^{-6}\ \mathrm{s})} = \boxed{251\ \mathrm{V}}$$

(b) An example of the ring shown in Figure 22.46b is the alternator in your car. If you were driving during a lightning storm and this large bolt of lightning hit at 50.0 m away it is possible to fry your alternator or your battery because of this large voltage surge. In addition, the hair at the back of your neck would stand on end because it would become statically charged.

22.103 (a) Using Equation 22.9, where the minus sign is from Lenz's law:

$$|\mathsf{E}| = L\frac{\Delta I}{\Delta t} = (25.0\ \mathrm{H})\frac{(100\ \mathrm{A})}{1.00\times10^{-6}\ \mathrm{s}} = \boxed{2.50\times10^9\ \mathrm{V}}$$

(b) The voltage is so extremely high that arcing would occur and the current would not be reduced so rapidly.

(c) It is not reasonable to shut off such a large current in such a large inductor in such an extremely short time.

ELECTROMAGNETIC WAVES (23)

CONCEPTUAL QUESTIONS

23.1 (a) Since the generator has its negative charges at the top, a positive test charge would feel a force upward, and therefore the electric field would also point upward.

(b) Since at $t = T/4$, the generator is oriented so that there is no charge distribution in the vertical direction, a positive test charge would not feel a force, and therefore there the electric field would be zero. (The electric field shown to the right of the generator is the motion away form the source as a function of time.)

(c) In this case, the generator has its negative charges at the bottom, so a positive test charge would feel a force downward, and therefore the electric field will point downward.

(d) Again, as in part (a), the generator has its negative charges at the top. Thus, a positive test charge will feel a force upward and the electric field will point upward.

23.4 In the left figure, the electric field is parallel to the wire enabling charges to move like in an antenna. In the right figure, the charges can only move the width of the wire.

23.7 If the DC voltages fluctuate or if the circuit is accelerated, electromagnetic waves could be emitted from a DC circuit.

23.10 The entire FM radio band lies between channels 6 and 7. The TV video signal is AM, while the TV audio is FM. It is possible to pick up the audio portion of Channel 6 on the low end of your FM radio receiver because the audio portion is FM and is broadcast at 88 MHz, which is in the range of an FM radio receiver.

23.13 A radio circuit resonates at the frequency it is receiving. That is how you can hear the radio station.

23.16 Yes, since visible frequencies are approximately 5×10^{14} Hz (from Example 23.3), they are substantially higher frequency than that of conventional electronic transmissions, so laser telephone transmissions should be able to carry substantially higher quantities of information than electronic transmissions. Since ELF radio communications with submarines are at extremely low frequencies, very little information can be transmitted because of the low frequencies.

23.19 High absorption means that the laser's energy is absorbed rapidly and does not penetrate very deeply. Hence the effect of the laser is localized. Since essentially all the energy of the laser is absorbed in the surface layers of the cornea, there will be very limited damage to the lens and retina because the energy isn't transmitted that deep.

PROBLEMS

23.1 From Section 21.9, we know that $\mu_0 = 4\pi \times 10^{-7}$ T·m/A, and from Section 18.5, we know that $\varepsilon_0 = 8.85 \times 10^{-12}$ F/m, so that Equation 23.1 becomes:

$$c = \frac{1}{\sqrt{\left(4\pi \times 10^{-7} \text{ T·m/A}\right)\left(8.8542 \times 10^{-12} \text{ F/m}\right)}} = \boxed{2.999 \times 10^8 \text{ m/s}}.$$

The units work as follows:

$$[c] = \frac{1}{\sqrt{\text{T·F/A}}} = \sqrt{\frac{A}{\text{T·F}}} = \sqrt{\frac{C/s}{(\text{N·s/C·m}) \times (C^2/J)}} = \sqrt{\frac{J·m}{N·s^2}} = \sqrt{\frac{(N·m)m}{N·s^2}} = \sqrt{\frac{m^2}{s^2}} = \text{m/s}.$$

23.7   Using Equation 23.3, $c = f\lambda$, we can solve for the frequency since we know the speed of light and are given the wavelength:

$$f = \frac{c}{\lambda} = \frac{2.998\times10^8 \text{ m/s}}{11.12 \text{ m}} = 2.696\times10^7 \text{ s}^{-1} = \boxed{26.96 \text{ MHz}}$$

23.13 From Equation 2.3, we know that $v = \dfrac{d}{t}$, and since we know the speed of light and the distance from the sun to the earth, we can calculate the time:

$$t = \frac{d}{c} = \frac{1.50\times10^{11} \text{ m}}{3.00\times10^8 \text{ m/s}} = \boxed{500 \text{ s}}$$

23.19 (a) Using Equation 23.3, we can calculate the frequency given the speed of light and the wavelength of the radiation:

$$f = \frac{c}{\lambda} = \frac{3.00\times10^8 \text{ m/s}}{193\times10^{-9} \text{ m}} = 1.55\times10^{15} \text{ s}^{-1} = \boxed{1.55\times10^{15} \text{ Hz}}$$

(b) The shortest wavelength of visible light is 380 nm, so that:

$$\frac{\lambda_{\text{visible}}}{\lambda_{\text{UV}}} = \frac{380 \text{ nm}}{193 \text{ nm}} = 1.97 .$$

In other words, the UV radiation is 97% more accurate than the shortest wavelength of visible light, or almost twice as accurate!

23.25 Using Equation 23.4b, we see that:

$$I = \frac{cB_0^2}{2\mu_0} = \frac{\left(3.00\times10^8 \text{ m/s}\right)\left(4.00\times10^{-9} \text{ T}\right)^2}{2\left(4\pi\times10^{-7} \text{ T}\cdot\text{m/A}\right)} = \boxed{1.91\times10^{-3} \text{ W/m}^2} .$$

The units work as follows:

$$[I] = \frac{(\text{m/s})\text{T}^2}{\text{T}\cdot\text{m/A}} = \frac{\text{T}\cdot\text{A}}{\text{s}} = \frac{(\text{N/A}\cdot\text{m})(\text{A})}{\text{s}} = \frac{\text{N}}{\text{s}\cdot\text{m}} = \frac{\text{J/m}}{\text{s}\cdot\text{m}} = \frac{\text{W}}{\text{m}^2}$$

23.31 (a) Using Equation 23.2, $\dfrac{E}{B} = c$, we can determine the maximum magnetic field strength given the maximum electric field strength:

$$B_0 = \frac{E_0}{c} = \frac{1.00\times10^{11} \text{ N/C}}{3.00\times10^8 \text{ m/s}} = \boxed{333 \text{ T}} ,$$

recalling that $1 \text{ V/m} = 1 \text{ N/C}$.

(b) Using Equation 23.4a, we can calculate the intensity without using the results from part (a):

$$I = \frac{c\varepsilon_0 E_0^2}{2} = \frac{\left(3.00\times10^8 \text{ m/s}\right)\left(8.85\times10^{-12} \text{ C}^2/\text{N}\cdot\text{m}^2\right)\left(1.00\times10^{11} \text{ N/C}\right)^2}{2} = \boxed{1.33\times10^{19} \text{ W/m}^2}$$

(c) Recalling Equation 16.3, we can get an expression for the power in terms of the intensity: $P = IA$, and from Equation 6.9, we can express the energy in terms of the power provided. Since we are told the time of the laser pulse, we can calculate the energy delivered to a 1.00 mm^2 area per pulse:

$$E = P\Delta t = IA\Delta t = \left(1.328\times10^{19} \text{ W/m}^2\right)\left(1.00 \text{ mm}^2\right)\left(\frac{1 \text{ m}}{1000 \text{ mm}}\right)^2 \left(1.00\times10^{-9} \text{ s}\right) = 1.33\times10^4 \text{ J} = \boxed{13.3 \text{ kJ}}$$

23.37 Using Equation 22.22, $f_0 = \dfrac{1}{2\pi\sqrt{LC}}$, we can find the capacitance in terms of the resonant frequency:

$C = \dfrac{1}{4\pi^2 L f_0^2}$. Substituting for the frequency, using Equation 23.3, gives:

$$C = \frac{\lambda^2}{4\pi^2 L c^2} \frac{(196 \text{ m})^2}{4\pi^2 (800\times10^{-6} \text{ H})(3.00\times10^8 \text{ m/s})^2} = 1.35\times10^{-11} \text{ F} = \boxed{13.5 \text{ pF}}\ .$$

The units work as follows: $[C] = \dfrac{\text{m}^2}{\text{H}(\text{m/s})^2} = \dfrac{\text{s}^2}{\text{H}} = \dfrac{\text{s}^2}{\Omega\cdot\text{s}} = \dfrac{\text{s}}{\Omega} = \dfrac{\text{s}}{\text{V/A}} = \dfrac{\text{A}\cdot\text{s}}{\text{V}} = \dfrac{\text{C}}{\text{V}} = \text{F}\ .$

23.43 (a) From Equation 16.3, we know:
$$I = \frac{P}{A} = \frac{5.00\times10^{-3} \text{ W}}{1.00\times10^{-6} \text{ m}^2} = \boxed{5.00\times10^3 \text{ W/m}^2}$$

(b) Using Equation 23.4a, $I = \dfrac{c\varepsilon_0 E_0^2}{2}$, we can solve for the maximum electric field:

$$E_0 = \sqrt{\frac{2I}{c\varepsilon_0}} = \sqrt{\frac{2(5.00\times10^3 \text{ W/m}^2)}{(3.00\times10^8 \text{ m/s})(8.85\times10^{-12} \text{ C}^2/\text{N}\cdot\text{m}^2)}} = 1.94\times10^3 \text{ N/C}\ .$$

So, using Equation 17.3, we can calculate the force on a 2.00 nC charge:
$$F = qE_0 = (2.00\times10^{-9} \text{ C})(1.94\times10^3 \text{ N/C}) = \boxed{3.88\times10^{-6} \text{ N}}\ .$$

(c) Using Equations 21.1 and 23.2, we can write the maximum magnetic force in terms of the electric field, since the electric and magnetic fields are related for electromagnetic radiation:

$$F_{B,\text{max}} = qvB_0 = \frac{qvE_0}{c} = \frac{(2.00\times10^9 \text{ C})(400 \text{ m/s})(1.94\times10^3 \text{ N/C})}{3.00\times10^8 \text{ m/s}} = \boxed{5.18\times10^{-12} \text{ N}}\ .$$

So the electric force is approximately 6 orders of magnitude stronger than the magnetic force.

23.49 (a) Using Equations 23.3, and 22.22, we can solve for the inductance: $f = \dfrac{c}{\lambda} = \dfrac{1}{2\pi\sqrt{LC}}$, so that

$$L = \frac{\lambda^2}{4\pi^2 C c^2} = \frac{(300\times10^{-9} \text{ m})^2}{4\pi^2 (1.00\times10^{-12} \text{ F})(3.00\times10^8 \text{ m/s})^2} = \boxed{2.53\times10^{-20} \text{ H}}$$

(b) This inductance is unreasonably small.

(c) The wavelength is too small.

(24) GEOMETRICAL OPTICS

CONCEPTUAL QUESTIONS

24.1 Powder makes the surface rough, scattering the reflected light in many different directions. This removes the shine and is called "diffusion".

24.4 It just proves that the speed of light is much faster than the speed of sound. A five seconds difference between the flash and boom equals approximately one mile. From this you could get a good estimate for the speed of sound. However, all you determine about the speed of light is that it is very, very fast in comparison.

24.7 The light ray is refracted as it moves from water to air. A person's eye extrapolates back along the current path of the light ray, and the legs appear short.

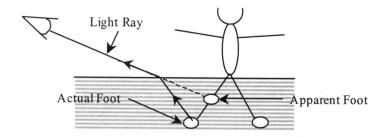

24.10 Looking at Equation 24.6a, $\frac{1}{d_o} + \frac{1}{d_i} = \frac{1}{f}$, if $f = \infty$, then $d_o = -d_i$, and the image forms on top of the original source.

24.13 No, the image exists whether or not there is a screen. You cannot see the image, however, unless it is projected on a screen.

24.16 A negative magnification means the image is upside down or "inverted". A magnification that is less than 1 in magnitude means the image is smaller than the object.

24.19 For the larger mirror, since the rays come out parallel, the filament is at the focal point for the mirror. For the smaller mirror, since the rays return the way they came in, the filament is at the radius of curvature, or at twice the focal length of the small mirror.

24.22 Looking at Equation 24.6a, $\frac{1}{d_o} + \frac{1}{d_i} = \frac{1}{f}$, if $f = \infty$, then $d_o = -d_i$, and the image formed is virtual, and the same distance behind the mirror as the image was in front of the mirror, and is the same height as the image and is upright.

24.25 Yes, place a convex lens beyond the eyepiece, which has a focal length that is smaller than the distance from the eyepiece to the virtual final image of the microscope. This will produce a real image, which can be projected. The closer the focal length is to the distance from the eyepiece to the virtual final image of the microscope, the larger the final real image will be.

24.28 Diffused because the light is scattered independent of wavelength.

24.31 If the curvature of a contact lens is not the same as the cornea the tear layer between the contact and cornea acts as a lens. If the tear layer is thinner in the center than at the edges, it has a negative power, for example. The effective power of the contact lens in the eye is the power of the dry contact lens plus the power of the tear layer.

24.34 For a fixed percent uncertainty, doubling the distance means doubling the absolute uncertainty. The more correction required, the more corneal surface that must be removed, which means a greater absolute uncertainty and the more likely the person is to have fuzzier vision after the operation.

24.37 The red object reflects only red light. Since there is no "red" in a green light, the object appears black. On a black background this means the object would seem to disappear because it appears the same color as the background.

24.40 If the vision tests are done in very dim light, the rods would be predominantly working, and that would provide a way to study their function alone.

PROBLEMS

24.1

From ray-tracing and the law of reflection, we know that the angle of incidence is equal to the angle of reflection, so the top of the mirror must extend to at least halfway between his eyes and the top of his head. And the bottom must go down to halfway between his eyes and the floor. This result independent of how far he stands from the wall. So,

$$a = \frac{0.13 \text{ m}}{2} = 0.065 \text{ m}, \quad b = \frac{1.65 \text{ m}}{2} = 0.825 \text{ m} \text{ and}$$

$$L = 1.65 \text{ m} + 0.13 \text{ m} - a - b = 1.78 \text{ m} - \frac{0.13 \text{ m}}{2} - \frac{1.65 \text{ m}}{2} = 0.89 \text{ m} .$$

So, the bottom is $b = \boxed{0.825 \text{ m}}$ from floor and the top is $b + L = 0.825 \text{ m} + 0.89 \text{ m} = \boxed{1.715 \text{ m}}$ from floor.

24.7 Using Equation 24.2, we find:

$$n = \frac{c}{v} = \frac{2.997 \times 10^8 \text{ m/s}}{2.012 \times 10^8 \text{ m/s}} = \boxed{1.490} .$$

From Table 24.1, the most likely substance is then $\boxed{\text{polystyrene}}$.

24.13 Using Equation 24.3, $n_1 \sin \theta_1 = n_2 \sin \theta_2$, we can solve for the unknown index of refraction:

$$n_2 = n_1 \frac{\sin \theta_1}{\sin \theta_2} = \frac{(1.333)(\sin 45.0°)}{\sin 40.3°} = \boxed{1.46} .$$

From Table 24.1, the most likely solid substance is $\boxed{\text{fused quartz}}$.

24.19 Using Equation 24.4, and the indices of refraction from Table 24.1 gives a critical angle of:

$$\theta_c = \sin^{-1}\left(\frac{n_2}{n_1}\right) = \sin^{-1}\left(\frac{1.52}{1.66}\right) = \boxed{66.3°} .$$

24.25 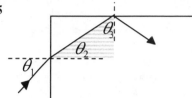 To do this problem, we need to work the problem from the inside backwards. First, we can determine the angle θ_3 from Equation 24.4, since that is the angle we want set to the critical angle for total internal reflection:

$$\theta_3 = \theta_c = \sin^{-1}\left(\frac{n_2}{n_1}\right) = \sin^{-1}\left(\frac{1.52}{1.66}\right) = 66.3°\,.$$

We can determine θ_2 from geometry. Looking at the shaded triangle in the figure, we see that:

$$90° + \theta_2 + \theta_3 = 180°,\text{ so that } \theta_2 = 90.0° - \theta_3 = 90.0° - 66.3° = 23.7°\,.$$

Finally, using Snell's Law:

$$n_1 \sin\theta_1 = n_2 \sin\theta_2\,,$$

we can determine the initial angle, where $n_1 = 1.00$ because outside the glass fiber is air:

$$\theta_1 = \sin^{-1}\left(\frac{n_2 \sin\theta_2}{n_1}\right) = \sin^{-1}\left[\frac{1.66\sin(23.7°)}{1.00}\right] = \boxed{41.9°}$$

24.31 Using Snell's Law (Equation 24.3), we have:

$$n_1 \sin\theta_1 = n_2 \sin\theta_2 \text{ and } n_1' \sin\theta_1' = n_2' \sin\theta_2'\,.$$

We can set θ_2 equal to θ_2', because the angles of refraction to be equal. So, combining the equations gives:

$$\frac{n_1 \sin\theta_1}{n_2} = \frac{n_1' \sin\theta_1'}{n_2'}\,.$$

We know that $n_1 = n_1' = 1.00$ because the light is entering from air. Finally, since we are given $\theta_1 = 55.0°$, and from Table 24.2, we find the 610 nm light in fused quartz has $n_2 = 1.456$ and the 470 nm light in flint glass has $n_2' = 1.684$. So, we can solve for the incident angle θ_1':

$$\theta_1' = \sin^{-1}\left(\frac{n_1 n_2'}{n_2 n_1'}\sin\theta_1\right) = \sin^{-1}\left[\frac{(1)(1.684)}{(1.456)(1)}\sin 55.0°\right] = \boxed{71.3°}$$

24.37 Using Equation 24.5, $P = \dfrac{1}{f}$, we can solve for the focal length of your eyeglasses, recalling that $1\,\text{D} = 1/\text{m}$:

$$f = \frac{1}{P} = \frac{1}{-4.50\,\text{D}} = -0.222\,\text{m} = \boxed{-22.2\,\text{cm}}\,.$$

24.43 Using Equation 24.5, we can solve for the focal length for your father's reading glasses:

$$f = \frac{1}{P} = \frac{1}{2.25\,\text{D}} = 0.444\,\text{m} = 44.4\,\text{cm}\,.$$

In order to burn a hole in the paper, you want to have the glasses exactly one focal length from the paper, so the glasses should be $\boxed{44.4\,\text{cm}}$ from the paper.

24.49 (a) Using Equation 24.6a, $\dfrac{1}{d_o} + \dfrac{1}{d_i} = \dfrac{1}{f}$, we can first determine the image distance:

$$d_i = \left(\frac{1}{f} - \frac{1}{d_o}\right)^{-1} = \left(\frac{1}{10.0 \text{ cm}} - \frac{1}{8.50 \text{ cm}}\right)^{-1} = -56.67 \text{ cm} .$$

Then, using Equation 24.6b, we can determine the magnification:

$$m = -\frac{d_i}{d_o} = \frac{56.67 \text{ cm}}{8.50 \text{ cm}} = \boxed{6.67} .$$

(b) Using Equations 24.6a and 24.6b again gives:

$$d_i = \left(\frac{1}{10.0 \text{ cm}} - \frac{1}{9.50 \text{ cm}}\right)^{-1} = -190 \text{ cm}$$

and a magnification of

$$m = -\frac{d_i}{d_o} = \frac{190 \text{ cm}}{9.5 \text{ cm}} = \boxed{+20.0} .$$

(c) The magnification, m, increases rapidly as you increase the object distance toward the focal length.

24.55 Using Equation 24.5, we can determine the power for the mirror:

$$P = \frac{1}{f} = \frac{1}{-3.00 \text{ m}} = \boxed{-0.333 \text{ D}}$$

24.61 *Step 1*: Image formation by a mirror is involved.
Step 2: Use the thin lens equations to solve this problem.
Step 3: Find: f.
Step 4: Given: $m = 1.50$, $d_o = 0.120 \text{ m}$.
Step 5: No ray tracing is needed.

Step 6: Using Equation 24.6b, $m = -\dfrac{d_i}{d_o}$, we know that

$$d_i = -md_o = -(1.50)(0.120 \text{ m}) = -0.180 \text{ m} .$$

Then, using Equation 24.6a, $\dfrac{1}{d_o} + \dfrac{1}{d_i} = \dfrac{1}{f}$, we can determine the focal length:

$$f = \left(\frac{1}{d_i} + \frac{1}{d_o}\right)^{-1} = \left(\frac{1}{-0.180 \text{ m}} + \frac{1}{0.120 \text{ m}}\right)^{-1} = \boxed{0.360 \text{ m}} .$$

Step 7: Since the focal length is larger than the object distance, we are dealing with **case 2**. For **case 2**, we should have a virtual image, a negative image distance and a positive (greater than one) magnification. Our answer is consistent with these expected properties, so it is reasonable.

24.67 Beginning with the result from Problem 24.52 and the fact that $R = -2f$ for a convex mirror, we have:

$$m = \frac{f}{f - d_o} = \frac{-R/2}{(-R/2) - d_o} = \frac{R}{R + 2d_o} .$$

As R gets smaller, the magnification gets smaller:

$$m \to \frac{0}{2d_o} = 0 .$$

24.73 Using Equation 24.9, $M = -\dfrac{f_o}{f_e}$, we can determine the focal length of the eyepiece since we know the magnification and the focal length of the objective:

$$f_e = -\frac{f_o}{M} = -\frac{75.0 \text{ cm}}{-7.50} = \boxed{+10.0 \text{ cm}}$$

24.79 Using the lens-to-retina distance of 2.00 cm and Equation 24.6a, we can determine the power at an object distance of 3.00 m:

$$P = \frac{1}{d_o} + \frac{1}{d_i} = \frac{1}{3.00 \text{ m}} + \frac{1}{0.0200 \text{ m}} = \boxed{+50.3 \text{ D}}$$

24.85 Since normal distant vision has a power of 50.0 D (see Example 24.13) and the laser vision correction reduced the power of her eye by 7.00 D, she originally had a power of 57.0 D. We can determine her original far point using Equation 24.6a, $P = \dfrac{1}{d_o} + \dfrac{1}{d_i}$, where the object distance is her far point and the image distance is the lens-to-retina distance:

$$d_o = \left(P - \frac{1}{d_i}\right)^{-1} = \left(57.0 \text{ D} - \frac{1}{0.0200 \text{ m}}\right)^{-1} = \boxed{0.143 \text{ m}} \,.$$

So, originally without corrective lenses, she could only see images 14.3 cm (or closer) to her eye.

24.91 From Example 24.13, we know that the normal power for distant vision is 50.0 D. So, for this woman, since she has a 10.0% ability to accommodate, her maximum power is:

$$P = (1.10)(50.0 \text{ D}) = 55.0 \text{ D} \,.$$

Thus, using Equation 24.6a, $P = \dfrac{1}{d_o} + \dfrac{1}{d_i}$ we can determine the nearest object she can see clearly since we know the image distance must be the lens-to-retina distance of 2.00 cm:

$$d_o = \left(P - \frac{1}{d_i}\right)^{-1} = \left(55.0 \text{ D} - \frac{1}{0.0200 \text{ m}}\right)^{-1} = 0.200 \text{ m} = \boxed{20.0 \text{ cm}}$$

24.97 When an object is held 25.0 cm from the person's eyes, the contact lens and tear layer must produce an image 29.0 cm away. Since the correction mechanism is contact lenses, the image distance is then 29.0 cm, and negative, so that $d_i = -0.290 \text{ m}$. The object distance is 25.0 cm and positive, so that $d_o = 0.250 \text{ m}$. Therefore, using Equation 24.6a, we can determine the power of the contact lens and tear layer:

$$P = \frac{1}{d_o} + \frac{1}{d_i} = \frac{1}{0.250 \text{ m}} + \frac{1}{-0.290 \text{ m}} = 0.552 \text{ D} \,.$$

So, since the contact lens has a power of +0.750 D, we can calculate the power of the tear layer:

$$P_{tear} = P - P_{lens} = 0.552 \text{ D} - (+0.750 \text{ D}) = \boxed{-0.198 \text{ D}} \,.$$

24.103 (a) When viewing an object infinitely far away, the contact lens must produce an image 1.00 m away. So, the object distance will be infinity, and the image distance is 1.00 m and negative, so that $d_i = -1.00$ m.

Using Equation 24.6a, we get a contact lens power of:

$$P = \frac{1}{d_o} + \frac{1}{d_i} = \frac{1}{\infty} + \frac{1}{d_i} = \frac{1}{d_i} = \frac{1}{-1.00 \text{ m}} = \boxed{-1.00 \text{ D}}$$

(b) First, we must determine the power of the eye without correction. Using Equation 24.6a, where the object distance is the far point and the image distance is the lens-to-retina distance (and positive), we get:

$$P = \frac{1}{d_o} + \frac{1}{d_i} = \frac{1}{1.00 \text{ m}} + \frac{1}{0.0200 \text{ m}} = 51.0 \text{ D}.$$

With accommodation, the maximum power this person has is 8.00% larger than that of the relaxed eye (the far vision power), so that:

$$P_{\text{max}} = (108.00\%)P = (1.0800)(51.0 \text{ D}) = 55.08 \text{ D}.$$

Now, the contact lenses will reduce the maximum power of the eye:

$$P_{\text{w/correction}} = P_{\text{max}} + P_{\text{lens}} = 55.08 \text{ D} + (-1.00 \text{ D}) = 54.08 \text{ D}.$$

Finally, we can determine the near point by again using Equation 24.6a, where the image distance is the lens-to-retina distance to focus the object on the retina:

$$d_o = \left(P_{\text{w/correction}} - \frac{1}{d_i} \right)^{-1} = \left(54.08 \text{ D} - \frac{1}{0.0200 \text{ m}} \right)^{-1} = 0.2451 \text{ m}.$$

Therefore, the closest object she will be able to see clearly with contacts on will be at a distance of $\boxed{24.5 \text{ cm}}$.

24.109 (a) From Chapter 13, we can get an expression for the heat transferred in terms of the mass of tissue ablated:

$$Q = mc\Delta T + mL_v = m(c\Delta T + L_v),$$

where the heat capacity is given in Table 13.1, $c = 4186$ J/kg·°C, and the latent heat of vaporization is given in Table 13.2, $L_v = 2256 \times 10^3$ J/kg. Solving for the mass gives:

$$m = \frac{Q}{c\Delta T + L_v} = \frac{0.500 \times 10^{-3} \text{ J}}{(4186 \text{ J/kg·°C})(100°C - 34.0°C) + 2256 \times 10^3 \text{ J/kg}} = 1.975 \times 10^{-10} \text{ kg}.$$

Now, since the corneal tissue has the same properties as water, its density is then 1000 kg/m³. Since we know the diameter of the spot, we can determine the thickness of the layer ablated: $\rho = \frac{m}{V} = \frac{m}{\pi r^2 t}$, so that:

$$t = \frac{m}{\pi r^2 \rho} = \frac{1.975 \times 10^{-10} \text{ kg}}{\pi (0.500 \times 10^{-3} \text{ m})^2 (1000 \text{ kg/m}^3)} = 2.515 \times 10^{-7} \text{ m} = \boxed{0.251 \text{ μm}}.$$

(b) Yes, this thickness implies that the shape of the cornea can be very finely controlled, producing normal distant vision in more than 90% of patients.

25 WAVE OPTICS

CONCEPTUAL QUESTIONS

25.1   Light exhibits interference and diffraction, which are wave characteristics indicating that light is a wave.

25.4   Light behaves like a wave when it exhibits interference and diffraction. Light behaves like a ray when it is polarized, transmitted, or reflected like when passing from air to water.

25.7   Yes, Huygen's principle applies to all types of waves like water, sound and light.

25.10   You could have a large single slit that would produce a minimum, but the maximum would not appear. Since $D\sin\theta = \lambda$, if D was just less than λ, there would be no maximum.

25.13   Yes, if the lines in a diffraction grating are too close together, then the diffraction would be for light in the ultraviolet, not the visible. (Smaller separation distances imply smaller wavelengths.)

25.16   In addition to pigmentation, the wing's color is greatly affected by constructive interference of certain wavelength reflected from its film-coated surface. In this case, the reflected wavelength is green light.

25.19   The fact that the bright spots are evenly spaced is a characteristic of a double slit. The fact that some of the bright spots are dim on either side of the center is a single slit characteristic, because the single slit pattern produces a minimum where the double slit would otherwise produce a maximum. The slit width is smaller than the separation between the slits because the central maximum for the single slit pattern has minimums within it suggesting that the double slit pattern is more closely space, and therefore the double slit separation is larger.

25.22   For constructive interference, two originally in-phase light waves must differ by an integral number of wavelengths. For destructive interference, the two originally in-phase light waves must differ by an integral number of wavelengths plus an additional half of a wavelength. This would be unaffected by reflection from a surface where the light reflects off a material of lower index of refraction, but a phase shift of a half a wavelength would be added when the light reflects off a material of higher index of refraction. In the case of refraction, there would be no phase shift, but it is necessary to measure the difference in distance in terms of the wavelength of the light in the current material.

25.25   When reflecting off a material with a HIGHER index of refraction, the reflection picks up an additional phase change. At the top of the glass slide, the light is reflecting in air off of glass, so there will be a phase change because the index of refraction of the glass is higher than that of air. At the bottom of the glass slide, the light is reflecting in glass off of water, so there will not be a phase change because the index of refraction of the water is less than that of glass. Finally, at the glass slide below the water, the light is reflecting in water off of glass, so there will be a phase change because the index of refraction of the glass is higher than that of the water.

25.28   Since a soap bubble has a higher index of refraction than the air on either side of it, in order to produce the same phase changes upon reflection, the index of refraction of the material used as a coating must be greater than the index of refraction of glass. Since the index of refraction of glass is relatively large, there are very few substances that could be used for our coating. Next, since the coating must be thin when compared to the wavelength of visible light, we must have a layer that is on the order of 100 atoms thick. In order to get a thin film of substances such as diamond, which has an index of refraction that we need, the lens would be subjected to a high temperature and vacuum, which would warp the lens surface. Therefore, it would be impractical. The coatings you would find on lenses today reduce glare by polarization, not by interference.

25.31 No, because sound waves are longitudinal not transverse waves. Polarization occurs in the plane perpendicular to the motion, therefore it effects only transverse waves.

25.34 This means that the larger the wavelength the less the scattering, since the amount of scattering is inversely proportional to λ^4. During the day, the sky appears blue because the blue light from the sun is scattered more and therefore we see it. The red light is not scattered as much, so it goes more in a straight line. Thus at sunrise or sunset when the path of the perceived sunlight is a long, straight path through the atmosphere, we see primarily reddish colors.

PROBLEMS

25.1 Using Equation 25.2, we can calculate the wavelength of light in water. The index of refraction for water is given in Table 24.1, so that:

$$\lambda_n = \frac{\lambda}{n} = \frac{\lambda}{1.333} = \boxed{0.750\lambda}.$$

So the wavelength of light in water is 0.750 times the wavelength in air.

25.7 Using Equation 25.3, $d \sin \theta = m\lambda$ for $m = 0,1,2,3,...$, we can calculate the angle for $m = 3$, given the wavelength and the slit separation:

$$\theta = \sin^{-1}\left(\frac{m\lambda}{d}\right) = \sin^{-1}\left[\frac{(3)(580 \times 10^{-9} \text{ m})}{0.100 \times 10^{-3} \text{ m}}\right] = \boxed{0.997°}$$

25.13 Looking at Equation 25.3, $d \sin \theta = m\lambda$, we notice that the highest order occurs when $\sin \theta = 1$, so the highest order is:

$$m = \frac{d}{\lambda} = \frac{25.0 \times 10^{-6} \text{ m}}{400 \times 10^{-9} \text{ m}} = 62.5.$$

Since m must be an integer, the highest order is then $\boxed{m = 62}$.

25.19 From Problem 25.18, we have an expression for the distance between fringes, so that:

$$\Delta y = \frac{x\lambda}{d} = \frac{(3.00 \text{ m})(633 \times 10^{-9} \text{ m})}{0.0800 \times 10^{-3} \text{ m}} = 2.37 \times 10^{-2} \text{ m} = \boxed{2.37 \text{ cm}}$$

25.25 The second order maximum is constructive interference, so we use Equation 25.5 for diffraction gratings, $d \sin \theta = m\lambda$ for $m = 0,1,2,3,...$, where the second order maximum has $m = 2$. Next, we need to determine the slit separation by using the fact that there are 5000 lines per centimeter:

$$d = \frac{1}{5000 \text{ slits/cm}} \times \frac{1 \text{ m}}{100 \text{ cm}} = 2.00 \times 10^{-6} \text{ m}.$$

So since $\theta = 45.0°$, we can determine the wavelength of the light:

$$\lambda = \frac{d \sin \theta}{m} = \frac{(2.00 \times 10^{-6} \text{ m})(\sin 45.0°)}{2} = 7.07 \times 10^{-7} \text{ m} = \boxed{707 \text{ nm}}$$

25.31 The largest possible second order occurs when $\sin\theta_2 = 1$. Using Equation 25.5, $d\sin\theta_m = m\lambda$, we see that the value for the slit separation and wavelength are the same for the first and second order maximums, so that:

$$d\sin\theta_1 = \lambda \text{ and } d\sin\theta_2 = 2\lambda,$$

so that:

$$\frac{\sin\theta_1}{\sin\theta_2} = \frac{1}{2}$$

Now, since we know the maximum value for $\sin\theta_2$, we can solve for the maximum value for θ_1:

$$\theta_1 = \sin\left(\frac{1}{2}\sin\theta_2\right)^{-1}$$

so that:

$$\theta_{1,max} = \sin\left(\frac{1}{2}\right)^{-1} = \boxed{30.0°}.$$

25.37 (a) First, we need to calculate the slit separation:

$$d = \frac{1 \text{ line}}{N} = \frac{1 \text{ line}}{30,000 \text{ lines/cm}} \times \frac{1 \text{ m}}{100 \text{ cm}} = 3.333 \times 10^{-7} \text{ m} = 333.3 \text{ nm}.$$

Next, using Equation 25.5, $d\sin\theta = m\lambda$, we see that the longest wavelength will be for $\sin\theta = 1$ and $m = 1$, so in that case, $d = \lambda = 333.3 \text{ nm}$, which is not visible.

(b) From part (a), we know that the longest wavelength is equal to the slit separation, or $\boxed{333 \text{ nm}}$.

(c) To get the largest number of lines/cm and still produce a complete spectrum, we want the smallest slit separation that allows the longest wavelength of visible light to produce a second order maximum, so $\lambda_{max} = 760 \text{ nm}$ (see Example 25.3). For there to be a second order spectrum, $m = 2$ and $\sin\theta = 1$, so

$$d = 2\lambda_{max} = 2(760 \text{ nm}) = 1.52 \times 10^{-6} \text{ m}.$$

Now, using the technique in step (a), only in reverse:

$$N = \frac{1 \text{ line}}{d} = \frac{1 \text{ line}}{1.52 \times 10^{-6} \text{ m}} \times \frac{1 \text{ m}}{100 \text{ cm}} = \boxed{6.58 \times 10^3 \text{ lines/cm}}.$$

25.43 Using Equation 25.6, $D\sin\theta = m\lambda$, where D is the slit width, we can determine the wavelength for the first minimum:

$$\lambda = \frac{D\sin\theta}{m} = \frac{(1.00 \times 10^{-6} \text{ m})(\sin 36.9°)}{1} = 6.004 \times 10^{-7} \text{ m} = \boxed{600 \text{ nm}}$$

25.49 The problem is asking us to find the ratio of D to d. For the single slit, using Equation 25.6, $D\sin\theta = n\lambda$, we have $n = 1$. For the double slit, using Equation 25.5 (because we have a maximum), $d\sin\theta = m\lambda$, we have $m = 5$. Dividing the single slit equation by the double slit equation, where the angle and wavelength are the same gives:

$$\frac{D}{d} = \frac{n}{m} = \frac{1}{5} \Rightarrow \boxed{\frac{D}{d} = 0.200}.$$

So, the slit separation is five times the slit width.

25.55 (a) Using Rayleigh's criterion, we can determine the angle (in radians) that is just resolvable:

$$\theta = 1.22\frac{\lambda}{D} = (1.22)\frac{(550\times10^{-9}\text{ m})}{3.00\times10^{-3}\text{ m}} = \boxed{2.24\times10^{-4}\text{ rad}}$$

(b) The distance s between two objects, a distance r away, separated by an angle θ is $s = r\theta$, so:

$$r = \frac{s}{\theta} = \frac{1.30\text{ m}}{2.237\times10^{-4}\text{ rad}} = 5.812\times10^{3}\text{ m} = \boxed{5.81\text{ km}}.$$

(c) Using the same equation as in part (b):

$$s = r\theta = (0.800\text{ m})(2.237\times10^{-4}\text{ rad}) = 1.789\times10^{-4}\text{ m} = \boxed{0.179\text{ mm}}$$

(d) Holding a ruler at arm's length, you can easily see the millimeter divisions; so you can resolve details 1.0 mm apart. Therefore, you probably can resolve details 0.2 mm apart at arm's length.

25.61 The minimum thickness will occur when there is one phase change, so for light incident perpendicularly, constructive interference first occurs when $2t = \dfrac{\lambda_n}{2} = \dfrac{\lambda}{2n}$. So, using the index of refraction for water from Table 24.1:

$$t = \frac{\lambda}{4n} = \frac{680\times10^{-9}\text{ m}}{(4)(1.33)} = 1.278\times10^{-7}\text{ m} = \boxed{128\text{ nm}}.$$

25.67 (a)

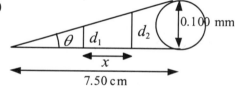

Two adjacent dark bands will have thicknesses differing by one wavelength, i.e. $\lambda = d_2 - d_1$, and $\tan\theta = \dfrac{\text{hair diameter}}{\text{slide length}}$, or

$$\theta = \tan^{-1}\left(\frac{0.100\times10^{-3}\text{ m}}{0.075\text{ m}}\right) = 0.076394°.$$

So since, $x\tan\theta = d_2 - d_1 = \lambda$, we see that

$$x = \frac{\lambda}{\tan\theta} = \frac{589\times10^{-9}\text{ m}}{\tan(0.076394°)} = 4.418\times10^{-4}\text{ m} = \boxed{0.442\text{ mm}}.$$

(b) The material makeup of the slides is irrelevant because it's the path difference *in the air* between the slides that gives rise to interference.

25.73 Using Equation 25.8, we can calculate the intensity:

$$I = I_0\cos^2\theta = (150\text{ W/m}^2)\cos^2(89.0°) = 4.57\times10^{-2}\text{ W/m}^2 = \boxed{45.7\text{ mW/m}^2}.$$

25.79 Using Equation 25.9, where n_2 is for crown glass and n_1 is for water (see Table 24.1), Brewster's angle is:

$$\theta_b = \tan^{-1}\left(\frac{n_2}{n_1}\right) = \tan^{-1}\left(\frac{1.52}{1.333}\right) = 48.75° = \boxed{48.8°}.$$

At 48.8° (Brewster's angle) the reflected light is completely polarized.

25.85 We're looking for the first minimum for single slit diffraction because the 50.0 m wide opening acts as a single slit. Using Equation 25.6, $D \sin \theta = m\lambda$, where $m = 1$, we can determine the angle for the first minimum:

$$\theta = \sin^{-1}\left(\frac{m\lambda}{D}\right) = \sin^{-1}\left[\frac{(1)(20.0 \text{ m})}{50.0 \text{ m}}\right] = 23.58° = \boxed{23.6°}.$$

Since the main peak for single slit diffraction is the main problem, a boat in the harbor at an angle greater than this first diffraction minimum will feel smaller waves. At the second minimum, the boat will not be affected by the waves at all:

$$\theta = \sin^{-1}\left(\frac{m\lambda}{D}\right) = \sin^{-1}\left[\frac{(2)(20.0 \text{ m})}{50.0 \text{ m}}\right] = 53.13° = \boxed{53.1°}.$$

25.91 (a) We use Equation 25.5 for diffraction gratings, $d \sin \theta = m\lambda$ for $m = 0, 1, 2, 3, \ldots$, where the fourth order maximum has $m = 4$. We first need to determine the slit separation by using the fact that there are 25,000 lines per centimeter:

$$d = \frac{1}{25,000 \text{ lines/cm}} \times \frac{1 \text{ m}}{100 \text{ cm}} = 4.00 \times 10^{-7} \text{ m}.$$

So since $\theta = 25.0°$, we can determine the wavelength of the light:

$$\lambda = \frac{d \sin \theta}{m} = \frac{\left(4.00 \times 10^{-7} \text{ m}\right)(\sin 25.0°)}{4} = 4.226 \times 10^{-8} \text{ m} = \boxed{42.3 \text{ nm}}$$

(b) This wavelength isn't in the visible spectrum.

(c) The number of slits in this diffraction grating is too large. Etching in integrated circuits can be done to a resolution of 50 nm, so slit separations of 400 nm is at the limit of what we can do today. This line spacing is too small to produce diffraction of visible light.

SPECIAL RELATIVITY

CONCEPTUAL QUESTIONS

26.1 Classical physics is not exactly correct at any velocity. At low velocities, however, it is a good enough approximation (to the accuracy of the measurements), so it gives the same results as experiment can measure. At higher velocities, clearly classical physics is incorrect.

26.4 Yes, if the airplane is not accelerating you are in an inertial reference frame. All inertial reference frames are equally valid; there is no preferred frame. If you choose to assume that the airplane is "stationary", there is no action you could perform on the plane (like pouring coffee or taking your pulse) that would prove otherwise.

26.7 The observer moving relative to the process will always measure a longer time than the observer at rest with the process because of time dilation. The observer who is at rest relative to the process measures proper time.

26.10 The astronaut does not notice the rate of his clock as slowing, or the length of his spaceship as shortening because he is at rest with the clock and the spaceship. According to the astronaut, the earthbound clocks run slow and the distance between stars that lie in lines parallel to his motion are shortened. Both the astronaut and the earthbound observers agree on his speed relative to the earth.

26.13 Yes, both Doppler effects say that the observed wavelength is larger for motion away. Since classical physics describes observed behavior at slow speeds, the classical Doppler effect must give the same results as the relativistic Doppler effect when considering slow speeds. (Remember, the classical Doppler effect isn't exactly correct, but it is a very good approximation for speeds significantly slower than the speed of light.)

26.16 Since relativity allows mass to be converted into energy and visa versa, mass and energy are not conserved separately (the classical case), but they are only conserved in total.

26.19 There is no upper limit on the momentum or the energy of an object with mass because both the momentum and energy are multiplied by the relativistic factor $\gamma = 1/\sqrt{1 - u^2/c^2}$, and as u approaches the speed of light, γ approaches infinity, giving no upper limit for parameters proportional to γ.

PROBLEMS

26.1 (a) Using the definition of gamma, where $v = 0.250c$:

$$\gamma = \left(1 - \frac{v^2}{c^2}\right)^{-1/2} = \left[1 - \frac{(0.250c)^2}{c^2}\right]^{-1/2} = \boxed{1.0328}$$

(b) Again, using the definition of gamma, now where $v = 0.500c$:

$$\gamma = \left[1 - \frac{(0.500c)^2}{c^2}\right]^{-1/2} = 1.1547 = \boxed{1.155}$$

Note: gamma is unitless and the results are reported to three digits difference from 1 in each case.

26.7 Using Equation 26.1 for time dilation,

$$\Delta t = \gamma \Delta t_0, \text{ where } \gamma = \left(1 - \frac{v^2}{c^2}\right)^{-1/2},$$

we see that:

$$\gamma = \frac{\Delta t}{\Delta t_0} = \frac{2065 \text{ s}}{900 \text{ s}} = 2.2944 = \left(1 - \frac{v^2}{c^2}\right)^{-1/2}.$$

Squaring the equation gives:

$$\gamma^2 = \left(\frac{c^2 - v^2}{c^2}\right)^{-1} = \frac{c^2}{c^2 - v^2}.$$

Cross-multiplying gives:

$$c^2 - v^2 = \frac{c^2}{\gamma^2},$$

and solving for the speed finally gives:

$$v = \sqrt{c^2 - \frac{c^2}{\gamma^2}} = c\left(1 - \frac{1}{\gamma^2}\right)^{1/2} = c\left[1 - \frac{1}{(2.2944)^2}\right]^{1/2} = 0.90003c = \boxed{0.900c}$$

26.13 Using the equations in Section 26.3 and the values given in Example 26.1:

(a) $L_0 = v\Delta t = (0.950c)(4.87 \times 10^{-6} \text{ s}) = (0.950)(2.998 \times 10^8 \text{ m/s})(4.87 \times 10^{-6} \text{ s}) = 1.387 \times 10^3 \text{ m} = \boxed{1.39 \text{ km}}$

(b) $L = v\Delta t_0 = (0.950)(2.998 \times 10^8 \text{ m/s})(1.52 \times 10^{-6} \text{ s}) = 432.9 \text{ m} = \boxed{0.433 \text{ km}}$

(c) Finally, using Equation 26.2: $L = \frac{L_0}{\gamma} = \frac{1.387 \times 10^3 \text{ m}}{3.20} = 433.4 \text{ m} = \boxed{0.433 \text{ km}}$.

26.19 Using Equation 26.3, we can add the relativistic velocities:

$$u = \frac{v + u'}{1 + (vu'/c^2)} = \frac{0.100c + 0.100c}{1 + [(0.100c)(0.100c)/c^2]} = \boxed{0.198c}$$

26.25 We are given: $u = 0.750c$ and $u' = 0.950c$. We want to find v . Starting with Equation 26.3,

$$u = \frac{v + u'}{1 + (vu'/c^2)},$$

we can solve for the speed. First, multiply both sides by the denominator:

$$u + v\frac{uu'}{c^2} = v + u'$$

and solving for the speed gives:

$$v = \frac{u' - u}{(uu'/c^2) - 1} = \frac{0.950c - 0.750c}{[(0.750c)(0.950c)/c^2] - 1} = \boxed{-0.696c} .$$

26.31 *Note all answers to this problem are reported to 5 significant figures, to distinguish the results.*

(a) We are given $v = -0.900c$ and $u = 0.990c$. Starting with Equation 26.3,

$$u = \frac{v + u'}{1 + \left(vu'/c^2\right)},$$

we now want to solve for u'. First, multiplying both sides by the denominator:

$$u + u'\frac{uv}{c^2} = v + u'$$

and solving for the probe's speed gives:

$$u' = \frac{u - v}{1 - \left(uv/c^2\right)} = \frac{0.990c - \left(-0.900c\right)}{1 - \left[\left(0.990c\right)\left(-0.900c\right)/c^2\right]} = \boxed{0.99947c}$$

(b) Assume it takes the probe a time t to reach the other galaxy. When the probe reaches the other galaxy, it will have traveled a distance (as seen on earth) of $d = x_0 + vt$ where $x_0 = 12 \times 10^9$ ly and $v = 0.900c$ because the galaxy is moving away from us. As seen from the earth, the probe is moving at a speed $u' = 0.9995c$, so the time it takes to travel that distance is:

$$t = \frac{d}{u'} = \frac{x_0 + vt}{u'}.$$

Now, we need to rewrite the equation so we can solve for the time. Multiplying by the denominator gives:
$u't = x_0 + vt$ and solving for t gives:

$$t = \frac{x_0}{u' - v} = \frac{12.0 \times 10^9 \text{ ly}}{0.99947c - 0.900c} \times \frac{\left(1 \text{ y}\right)c}{1 \text{ ly}} = \boxed{1.2064 \times 10^{11} \text{ y}}$$

(c) The radio signal travels at the speed of light, so the return time, t', is given by: $t' = \dfrac{d}{c} = \dfrac{x_0 + vt}{c}$, assuming the signal is transmitted as soon as the probe reaches the other galaxy. So, using the numbers, we can determine the time:

$$t' = \frac{12.0 \times 10^9 \text{ ly} + \left(0.900c\right)\left(1.2064 \times 10^{11} \text{ y}\right)}{c} = \boxed{1.2058 \times 10^{11} \text{ y}}$$

26.37 Beginning with Equation 26.5, $p = \gamma mu = \dfrac{mu}{\left[1 - \left(u^2/c^2\right)\right]^{1/2}}$, we can solve for the speed u. First, cross-multiply and square both sides, giving:

$$1 - \frac{u^2}{c^2} = \frac{m^2}{p^2}u^2.$$

Then, solving for u^2 gives:

$$u^2 = \frac{1}{\left(m^2/p^2\right) + \left(1/c^2\right)} = \frac{p^2}{m^2 + \left(p^2/c^2\right)}.$$

Finally, taking the square root gives: $u = \dfrac{p}{\sqrt{m^2 + \left(p^2/c^2\right)}}$. So, taking the values for the mass of the electron and the speed of light to five significant figures from the front cover of the book gives:

$$u = \frac{3.34 \times 10^{-21} \text{ kg} \cdot \text{m/s}}{\left\{\left(9.1094 \times 10^{-31} \text{ kg}\right)^2 + \left[\left(3.34 \times 10^{-21} \text{ kg} \cdot \text{m/s}\right)/\left(2.9979 \times 10^8 \text{ m/s}\right)\right]^2\right\}^{1/2}} = \boxed{2.988 \times 10^8 \text{ m/s}}$$

26.43 (a) From Table 6.1, we know the energy released from the nuclear fission of 1.00 kg of uranium is $\Delta E = 8.0 \times 10^{13}$ J. So, using Equation 26.7, $\Delta E = E_{\text{released}} = \Delta mc^2$, we can calculate the mass of uranium converted to energy:

$$\Delta m = \frac{\Delta E}{c^2} = \frac{8.0 \times 10^{13} \text{ J}}{\left(3.00 \times 10^8 \text{ m/s}\right)^2} = 8.89 \times 10^{-4} \text{ kg} = \boxed{0.89 \text{ g}}.$$

Note that since $1 \text{ J} = 1 \text{ kg} \cdot \text{m}^2/\text{s}^2$, the units work out correctly.

(b) To calculate the ratio, simply divide by the original mass:

$$\frac{\Delta m}{m} = \frac{8.89 \times 10^{-4} \text{ kg}}{1.00 \text{ kg}} = 8.89 \times 10^{-4} = \boxed{8.9 \times 10^{-4}}$$

26.49 Using Equation 26.8, we can determine the kinetic energy of the muon by determining the missing mass:

$$\text{KE}_{\text{rel}} = \Delta mc^2 = \left(m_\pi - m_\mu\right)c^2 = \left(\gamma - 1\right)m_\mu c^2.$$

So, solving for gamma will give us a way of calculating the speed of the muon. From the equation above, we see that:

$$\gamma = \frac{m_\pi - m_\mu}{m_\mu} + 1 = \frac{m_\pi - m_\mu - m_\mu}{m_\mu} = \frac{m_\pi}{m_\mu} = \frac{139.6 \text{ MeV}}{105.7 \text{ MeV}} = 1.32072.$$

Now, using the definition of gamma from Equation 26.1,

$$\gamma = \left[1 - \left(v^2/c^2\right)\right]^{-1/2},$$

we can solve for the speed of the muon. First, we square the equation and cross-multiply to get:

$$1 - \frac{v^2}{c^2} = \frac{1}{\gamma^2},$$

so solving for the speed gives:

$$v = c\sqrt{1 - \frac{1}{\gamma^2}} = c\sqrt{1 - \frac{1}{\left(1.32072\right)^2}} = \boxed{0.6532c}.$$

26.55 From Problem 26.7, we know that $\gamma = 2.2944$, so using Equation 26.8, we can determine the kinetic energy of the neutron:

$$\text{KE}_{\text{rel}} = \left(\gamma - 1\right)mc^2 = \left(2.2944 - 1\right)\left(939.6 \text{ MeV}\right) = \boxed{1216 \text{ MeV}}$$

26.61 (a) Using Equation 26.7, we can calculate the rest mass energy of a 1.00 kg mass. This rest mass energy is the energy released by the destruction of that amount of mass:

$$E_{\text{released}} = mc^2 = \left(1.00 \text{ kg}\right)\left(2.998 \times 10^8 \text{ m/s}\right)^2 = 8.988 \times 10^{16} \text{ J} = \boxed{8.99 \times 10^{16} \text{ J}}$$

(b) Using Equation 6.4, $\text{PE} = mgh$, we can determine how much mass can be raised to a height of 10.0 km:

$$m = \frac{\text{PE}}{gh} = \frac{8.99 \times 10^{16} \text{ J}}{\left(9.80 \text{ m/s}^2\right)\left(10.0 \times 10^3 \text{ m}\right)} = \boxed{9.17 \times 10^{11} \text{ kg}}$$

26.67 (a) Using Equation 26.1, $\Delta t = \gamma \Delta t_0$, we can solve for gamma:

$$\gamma = \frac{\Delta t}{\Delta t_0} = \frac{23.9 \text{ h}}{24.0 \text{ h}} = 0.9958 = \boxed{0.996}$$

(b) $\gamma < 1$ is not physical result. It must be ≥ 1.

(c) The earthbound observer must measure a *longer* time than the ship bound observer, so it is unreasonable to have the time shorter on earth than on the ship.

27 INTRODUCTION TO QUANTUM MECHANICS

CONCEPTUAL QUESTIONS

27.1 Population is a physical entity that is quantized. In other words, people come in units of one. You can't have a half of a person. So, populations are in discrete numbers and cannot have fractional value.

27.4 Consider a swing, or even better a merry-go-round. The child's angular momentum is quantized, meaning that he will spin only at certain speeds. For instance, if a child is moving at the slowest speed allowed, and he asks to go faster, you must push the child two, three, or four times faster in order to be in an allowed state. Since in our world Planck's constant is so small, this slowest speed is so small as to not be noticeable.

27.7 No, the photoelectric effect is the name given to the process in which light or other EM radiation ejects electrons from a material.

27.10 Charges in metals are free charges, in other words they are not tightly bound to a particular nucleus. This means that they are easier to remove than bound charges. So, the free charges that make metals good conductors are the reason that metals have lower binding energies than insulators because those free charges are less tightly bound to their nuclei. Insulators have no free charges, so their charges are bound to particular nuclei and cannot so easily be removed.

27.13 Ionizing radiation kills bacteria. Instead of UV light, one could use gamma rays and x-rays, which are more penetrating and deliver sufficient energy to kill microorganisms with one photon.

27.16 Wearing sunglasses that lack UV blockers increases your risk of UV damage to your eyes. The sunglasses will cause your eyes to dilate due to a decrease in visible light. Since the sunglasses do not block UV, when your eyes dilate, they are exposed to more UV radiation than they should be increasing the risk of UV damage. If the sunglasses block UV, then even though your eyes dilate, UV radiation is not allowed to penetrate the sunglasses and will not cause damage to your eyes.

27.19 Yes, technically it would be more correct to say $hf = qV + \text{BE}$, but since the binding energy is so much smaller than qV, it is a negligible term and is not included.

27.22 We don't feel the momentum of sunlight because it is really small, even on a hot day.

27.25 The photoelectric effect is evidence of the particle nature of EM radiation.

PROBLEMS

27.1 (a) Using Equation 27.3, we can calculate the difference in energy levels given the frequency of the radiation:

$$\Delta E = hf = \left(6.626\times10^{-34}\ \text{J}\cdot\text{s}\right)\left(1.7\times10^{13}\ \text{s}^{-1}\right) = 1.126\times10^{-20}\ \text{J}\ .$$

Then, using the definition of the electron volt, we can convert the units to electron volts:

$$\left(1.126\times10^{-20}\ \text{J}\right)\!\left(\frac{1\ \text{eV}}{1.602\times10^{-19}\ \text{J}}\right) = 7.03\times10^{-2}\ \text{eV} = \boxed{7.0\times10^{-2}\ \text{eV}}$$

(b) Using Equation 27.1, $E = nhf$, we can solve for n:

$$n = \frac{E}{hf} = \frac{\left(1.0\ \text{eV}\right)\left(1.602\times10^{-19}\ \text{J/eV}\right)}{\left(6.626\times10^{-34}\ \text{J}\cdot\text{s}\right)\left(1.7\times10^{13}\ \text{s}^{-1}\right)} = 14.22 = \boxed{14}\ .$$

27.7 The longest wavelength corresponds to the shortest frequency, or the smallest energy. Therefore, the smallest energy is when the kinetic energy is zero, so from Equation 27.5, $\text{KE} = hf - \text{BE} = 0$, we can calculate the binding energy (using Equation 23.3 to write the frequency in terms of the wavelength): $hf = \text{BE} = \dfrac{hc}{\lambda}$, so that:

$$\text{BE} = \frac{\left(6.626\times10^{-34}\ \text{J}\cdot\text{s}\right)\left(2.998\times10^{8}\ \text{m/s}\right)}{304\times10^{-9}\ \text{m}} = 6.534\times10^{-19}\ \text{J}\times\!\left(\frac{1.000\ \text{eV}}{1.602\times10^{-19}\ \text{J}}\right) = \boxed{4.08\ \text{eV}}$$

27.13 Using Equation 27.5, $\text{KE} = hf - \text{BE}$, and Equation 23.3 $c = \lambda f$, we see that:

$$hf = \frac{hc}{\lambda} = \text{BE} + \text{KE}\ ,$$

so that we can calculate the wavelength of the photons in terms of the energies:

$$\lambda = \frac{hc}{\text{BE} + \text{KE}} = \frac{\left(6.626\times10^{-34}\ \text{J}\cdot\text{s}\right)\left(2.998\times10^{8}\ \text{m/s}\right)}{2.24\ \text{eV} + 0.100\ \text{eV}}\times\!\left(\frac{1.000\ \text{eV}}{1.602\times10^{-19}\ \text{J}}\right) = 5.299\times10^{-7}\ \text{m} = \boxed{530\ \text{nm}}\ .$$

$\boxed{\text{Yes}}$, these photons are visible.

27.19 (a) Using Equation 27.4, we can determine the energy of the photons:

$$E = hf = \left(6.626\times10^{-34}\ \text{J}\cdot\text{s}\right)\left(90.0\times10^{6}\ \text{s}^{-1}\right) = 5.963\times10^{-26}\ \text{J} = \boxed{5.96\times10^{-26}\ \text{J}}\ .$$

Using the definition of an electron volt, we can convert the units to eV:

$$\left(5.963\times10^{-26}\ \text{J}\right)\!\left(\frac{1\ \text{eV}}{1.602\times10^{-19}\ \text{J}}\right) = \boxed{3.72\times10^{-7}\ \text{eV}}$$

(b) This implies that a tremendous number of photons must be broadcast per second! In order to have a broadcast power of, say 50.0 kW, it would take

$$\frac{50.0\times10^{3}\ \text{J/s}}{5.967\times10^{-26}\ \text{J/photon}} = 8.38\times10^{29}\ \text{photons/s}\ .$$

27.25 Using the conversion for Joules to electron volts and meters to nanometers gives:

$$hc = \left(6.626 \times 10^{-34} \text{ J} \cdot \text{s}\right)\left(2.998 \times 10^{8} \text{ m/s}\right)\left(\frac{10^{9} \text{ nm}}{1 \text{ m}}\right)\left(\frac{1.000 \text{ eV}}{1.602 \times 10^{-19} \text{ J}}\right) = 1239.997 \text{ eV} \cdot \text{nm} = \boxed{1240 \text{ eV} \cdot \text{nm}}$$

27.31 (a) Using Equation 27.4, we can first calculate the energy of each photon:

$$E_\gamma = hf = \left(6.626 \times 10^{-34} \text{ J} \cdot \text{s}\right)\left(650 \times 10^{3} \text{ s}^{-1}\right) = 4.307 \times 10^{-28} \text{ J} .$$

Then, using the fact that the broadcast power is 50.0 kW, we can calculate the number of photons per second:

$$N = \frac{50.0 \times 10^{3} \text{ J/s}}{4.307 \times 10^{-28} \text{ J/photon}} = 1.161 \times 10^{32} \text{ photons/s} = \boxed{1.16 \times 10^{32} \text{ photons/s}}$$

(b) To calculate the flux of photons, we assume the broadcast is uniform in all directions, so the area is the surface area of a sphere giving:

$$\Phi_N = \frac{N}{4\pi r^2} = \frac{1.161 \times 10^{32} \text{ photons/s}}{4\pi \left(100 \times 10^{3} \text{ m}\right)^2} = \boxed{9.24 \times 10^{20} \text{ photons/s} \cdot \text{m}^2}$$

27.37 (a) Using Equation 27.6, $p = \dfrac{h}{\lambda}$, we can solve for the wavelength of the photon:

$$\lambda = \frac{h}{p} = \frac{6.626 \times 10^{-34} \text{ J} \cdot \text{s}}{5.00 \times 10^{-27} \text{ kg} \cdot \text{m/s}} = 1.325 \times 10^{-7} \text{ m} = \boxed{133 \text{ nm}}$$

(b) Using Equation 27.7, $p = \dfrac{E}{c}$, we can solve for the energy and then convert the units to electron volts:

$$E = pc = \left(5.00 \times 10^{-27} \text{ kg} \cdot \text{m/s}\right)\left(2.998 \times 10^{8} \text{ m/s}\right) = 1.499 \times 10^{-18} \text{ J} \times \left(\frac{1 \text{ eV}}{1.602 \times 10^{-19} \text{ J}}\right) = \boxed{9.36 \text{ eV}}$$

27.43 Beginning with the two given equations: $E = \gamma mc^2$ and $p = \gamma mu$ gives:

$$\frac{E}{p} = \frac{\gamma mc^2}{\gamma mu} = \frac{c^2}{u} .$$

As the mass of a particle approaches zero, its velocity, u, will approach c, so that the ratio of energy to momentum approaches:

$$\lim_{m \to 0} \frac{E}{p} = \frac{c^2}{c} = c ,$$

which is consistent with Equation 27.7 for photons.

27.49 (a) Using Equations 27.6 and 7.1 we can calculate the wavelength of the neutron:

$$\lambda = \frac{h}{p} = \frac{h}{mv} = \frac{6.626 \times 10^{-34} \text{ J} \cdot \text{s}}{\left(1.675 \times 10^{-27} \text{ kg}\right)\left(1.00 \text{ m/s}\right)} = 3.956 \times 10^{-7} \text{ m} = \boxed{396 \text{ nm}} .$$

(b) Using Equation 6.3 we can calculate the kinetic energy of the neutron:

$$\text{KE} = \frac{1}{2}mv^2 = \frac{1}{2}\left(1.675 \times 10^{-27} \text{ kg}\right)\left(1.00 \text{ m/s}\right)^2 = 8.375 \times 10^{-28} \text{ J} \times \frac{1.00 \text{ eV}}{1.602 \times 10^{-19} \text{ J}} = \boxed{5.23 \times 10^{-9} \text{ eV}}$$

27.55 Using Equation 26.8, we can calculate the value for gamma: $KE = (\gamma - 1)mc^2 = \gamma mc^2 - mc^2$, so that

$$\gamma = \frac{KE + mc^2}{mc^2} = \frac{0.100 \text{ MeV} + 0.511 \text{ MeV}}{0.511 \text{ MeV}} = 1.1957 .$$

Then, using the definition for gamma, $\gamma = \left[1 - \left(u^2/c^2\right)\right]^{-1/2}$, we can solve for the speed:

$$u = c\left(1 - \frac{1}{\gamma^2}\right)^{1/2} = c\left[1 - \frac{1}{\left(1.1957\right)^2}\right]^{1/2} = 0.5482c .$$

Finally, using Equation 27.6, $p = h/\lambda$, and Equation 26.5, $p = \gamma mu$, we can determine the wavelength:

$$\lambda = \frac{h}{p} = \frac{h}{\gamma mu} = \frac{6.626 \times 10^{-34} \text{ J} \cdot \text{s}}{\left(1.1957\right)\left(9.109 \times 10^{-31} \text{ kg}\right)\left(0.5482\right)\left(2.998 \times 10^8 \text{ m/s}\right)} = \boxed{3.70 \times 10^{-12} \text{ m}}$$

27.61 Using Equation 27.10, $\Delta E\, \Delta t \geq \dfrac{h}{4\pi}$, we can determine the minimum uncertainty for its energy:

$$\Delta E \geq \frac{h}{4\pi\, \Delta t} = \frac{6.626 \times 10^{-34} \text{ J} \cdot \text{s}}{4\pi\left(3.00 \times 10^{-3} \text{ s}\right)} = 1.758 \times 10^{-32} \text{ J} \times \left(\frac{1 \text{ eV}}{1.602 \times 10^{-19} \text{ J}}\right) = \boxed{1.10 \times 10^{-13} \text{ eV}}$$

27.67 Using Equation 25.3, $d \sin\theta = m\lambda$, $m = 0,\ 1,\ 2,\ \ldots$, we can solve for the slit separation when $m = 1$ for the first-order maximum:

$$d = \frac{m\lambda}{\sin\theta} = \frac{(1)(0.167 \text{ nm})}{\sin 25.0°} = \boxed{0.395 \text{ nm}} .$$

27.73 (a) Using Equation 26.6, $E = \gamma mc^2$, we can find gamma for the 1.00 TeV proton:

$$\gamma = \frac{E}{mc^2} = \frac{\left(1.00 \times 10^{12} \text{ eV}\right)\left(1.602 \times 10^{-19} \text{ J/eV}\right)}{\left(1.673 \times 10^{-27} \text{ kg}\right)\left(2.998 \times 10^8 \text{ m/s}\right)^2} = 1065 = \boxed{1.06 \times 10^3}$$

(b) Using the using the definition for gamma, $\gamma = \left[1 - \left(u^2/c^2\right)\right]^{-1/2}$, we can solve for the speed:

$$u = c\left(1 - \frac{1}{\gamma^2}\right)^{1/2} = c\left[1 - \frac{1}{\left(1.065 \times 10^3\right)^2}\right]^{1/2} = 0.9999996c \quad c .$$

Finally, using Equation 26.5, where $u = c$ (to 4 significant figures):

$$p = \gamma mc = \left(1.065 \times 10^3\right)\left(1.673 \times 10^{-27} \text{ kg}\right)\left(2.998 \times 10^8 \text{ m/s}\right) = \boxed{5.34 \times 10^{-16} \text{ kg} \cdot \text{m/s}} .$$

(c) Using Equation 27.6, $p = h/\lambda$, we can calculate the proton's wavelength:

$$\lambda = \frac{h}{p} = \frac{6.626 \times 10^{-34} \text{ kg} \cdot \text{m/s}}{5.344 \times 10^{-16} \text{ kg} \cdot \text{m/s}} = 1.240 \times 10^{-18} \text{ m} = \boxed{0.00124 \text{ fm}}$$

27.79 First, we know the amount of heat absorbed by 1.00 kg of tissue is equal to the number of photons times the energy they each carry, so:

$$Q = NE_\gamma = \left(4.00 \times 10^{13}\right)\left(200 \times 10^3 \text{ eV}\right)\left(\frac{1.602 \times 10^{-19} \text{ J}}{\text{eV}}\right) = 1.282 \text{ J}.$$

Next, using Equation 13.2, $Q = mc\Delta T$, we can determine how much 1.00 kg of tissue is heated:

$$\Delta T = \frac{Q}{mc} = \frac{1.282 \text{ J}}{\left(1.00 \text{ kg}\right)\left(0.830 \text{ kcal/kg} \cdot ^\circ\text{C}\right)\left(4186 \text{ J/kcal}\right)} = \boxed{3.69 \times 10^{-4} \, ^\circ\text{C}}$$

27.85 (a) We want to use Equation 27.5, $KE = hf - BE$, to determine the binding energy, so we first need to determine an expression for hf. Using Equation 25.1, we know:

$$hf = \frac{hc}{\lambda} = \frac{\left(6.626 \times 10^{-34} \text{ J} \cdot \text{s}\right)\left(2.998 \times 10^8 \text{ m/s}\right)}{400 \times 10^{-9} \text{ m}} = \left(4.966 \times 10^{-19} \text{ J}\right)\left(\frac{1 \text{ eV}}{1.602 \times 10^{-19} \text{ J}}\right) = 3.100 \text{ eV}$$

And, since $KE = hf - BE$:

$$BE = hf - KE = 3.100 \text{ eV} - 4.00 \text{ eV} = \boxed{-0.90 \text{ eV}}$$

(b) The binding energy cannot be negative.

(c) The electron's kinetic energy is too large for the given photon energy; it cannot be greater than the photon energy.

ATOMIC PHYSICS

28

CONCEPTUAL QUESTIONS

28.1 By the beginning of the 19th century, an important fact was well established – the masses of reactants in specific chemical reactions always had a particular mass ratio. This is very strong indirect evidence for the existence of atoms. The first direct observation of atoms was in Brownian motion, the statistical fluctuation of particles caused by collisions with atoms. Now, it has become possible to accelerate ions and to detect them individually as well as measure their masses.

28.4 (c), J. J. Thomson measured the charge to mass ratio for the electron and then the Millikan oil drop experiment gave the charge of the electron, leading to a calculation of the mass of the electron.

28.7 Discrete spectra means that atomic and molecular spectra produce lines at certain angles and that some angles never produce spectral lines. Discrete spectra imply quantization of energy states in atoms and molecules because spectral lines come from electrons changing energy levels in an atom or molecule.

28.10 If a nucleus has Z protons and only one electron, that atom is called hydrogen-like. The energies and radii are related to the hydrogen atom by Equations 28.10 and 28.8 respectively.

28.13 n is the principal quantum number that labels the basic states of a system; ℓ is the angular momentum quantum number; m_ℓ is the angular momentum projection quantum number; s is the spin quantum number; and m_s is the spin projection quantum number.

28.16 For an electron, $s = 1/2$ and $m_s = \pm 1/2$. Protons and neutrons, like electrons, have $s = 1/2$, whereas photons have $s = 1$, other particles called pions have $s = 0$, etc.

28.19 Strontium and calcium have similar chemistry because they are both in the IIA group. They both have a filled s-shell in their ground state. Other elements that would have similar chemistry are: beryllium, magnesium, barium and radium.

28.22 Characteristic x-rays are produced when an electron drops into an inner-shell vacancy. The energy of characteristic x-rays becomes increasingly greater for heavier and heavier atoms because the energy levels for the heavier atoms are farther apart (see Equation 28.10), therefore producing more energy when an electron drops into an inner-shell.

28.25 The shorter wavelength of the blue light laser allows it to be focused more tightly and therefore burn smaller pits into the CD, which means it can store more information.

28.28 Fluorescence is any process in which an atom or molecule is excited by one type of energy and deexcites by emission of a different form of energy. Some states live much longer than others and are termed metastable. Phosphorescence is the deexcitation of a metastable state.

28.31 Population inversion is when the majority of the atoms are in a metastable state. Once population inversion is achieved, an electron spontaneously falls from the metastable state, emitting a photon. This photon finds another atom in the metastable state and stimulates it to decay, emitting a second photon of the same wavelength and in phase with the first. This is called stimulated emission and is the process by which a laser works. The probability of absorption of a photon is the same as the probability of stimulated emission, and so a majority of atoms must be in the metastable state to produce energy.

PROBLEMS

28.1 We can calculate the ratio of the masses by taking the ratio of the charge to mass ratios given:

$$\frac{q}{m_e} = 1.76 \times 10^{11} \ \text{C/kg} \ \text{and} \ \frac{q}{m_p} = 9.57 \times 10^7 \ \text{C/kg} ,$$

so that

$$\frac{m_p}{m_e} = \frac{q/m_e}{q/m_p} = \frac{1.76 \times 10^{11} \ \text{C/kg}}{9.57 \times 10^7 \ \text{C/kg}} = 1839 = \boxed{1.84 \times 10^3} .$$

The actual mass ratio is:

$$\frac{m_p}{m_e} = \frac{1.6726 \times 10^{-27} \ \text{kg}}{9.1094 \times 10^{-31} \ \text{kg}} = 1836 = 1.84 \times 10^3 ,$$

so to three digits, the mass ratio is correct.

28.7 (a) First, we need to calculate the surface area of the pollen grain, then divide that by the surface area of one side of the atom cube:

$$A_{\text{pollen}} = (6)(10^{-6} \ \text{m})^2 = 6 \times 10^{-12} \ \text{m}^2 , \ \text{and} \ A_{\text{atom}}/\text{side} = (10^{-10})^2 = 10^{-20} \ \text{m}^2 ,$$

so that:

$$\# \text{ atoms/pollen grain} = \frac{6 \times 10^{-12} \ \text{m}^2}{(10^{-10} \ \text{m})^2} = \boxed{6 \times 10^8}$$

(b) The uncertainty in the number is then given by the square root of the number divided by the number:

$$\Delta \# = \frac{(6 \times 10^8)^{1/2}}{6 \times 10^8} = \boxed{4 \times 10^{-5}}$$

28.13 Using Equation 28.12, $E_n = \dfrac{-13.6 \ \text{eV}}{n^2}$, we can determine the value for n, given the ionization energy:

$$n = \sqrt{\frac{-13.6 \ \text{eV}}{E_n}} = \left(\frac{-13.6 \ \text{eV}}{-0.85 \ \text{eV}}\right)^{1/2} = 4.0 = \boxed{4} .$$

(Remember than n must be an integer.)

28.19 (a) From page 607, we know that the UV range is from $\lambda = 10$ nm to approximately $\lambda = 380$ nm. Using

Equation 28.4, $\dfrac{1}{\lambda} = R\left(\dfrac{1}{n_f^2} - \dfrac{1}{n_i^2}\right)$, where $n_f = 2$ for the Balmer series, we can solve for n_i. Finding a

common denominator gives:

$$\frac{1}{\lambda R} = \frac{n_i^2 - n_f^2}{n_i^2 n_f^2},$$

so that $n_i^2 n_f^2 = \lambda R\left(n_i^2 - n_f^2\right)$, or $n_i = n_f\sqrt{\dfrac{\lambda R}{\lambda R - n_f^2}}$. The first line will be for the lowest energy photon, and

therefore the largest wavelength, so setting $\lambda = 380$ nm gives:

$$n_i = 2\sqrt{\frac{\left(380\times 10^{-9}\text{ m}\right)\left(1.097\times 10^{7}\text{ m}^{-1}\right)}{\left(380\times 10^{-9}\text{ m}\right)\left(1.097\times 10^{7}\text{ m}^{-1}\right) - 4}} = 9.94 \Rightarrow \boxed{n_i = 10}\ \text{ will be the first.}$$

(b) So, setting $\lambda = 760$ nm allows us to calculate the smallest value for n_i in the visible range:

$$n_i = 2\sqrt{\frac{\left(760\times 10^{-9}\text{ m}\right)\left(1.097\times 10^{7}\text{ m}^{-1}\right)}{\left(760\times 10^{-9}\text{ m}\right)\left(1.097\times 10^{7}\text{ m}^{-1}\right) - 4}} = 2.77 \Rightarrow n_i = 3.$$

So $n_i = 3$ to 9 are visible, or $\boxed{7 \text{ lines are in the visible}}$

(c) The smallest λ in the Balmer series would be for $n_i = \infty$, which corresponds to a value of:

$$\frac{1}{\lambda} = R\left(\frac{1}{n_f^2} - \frac{1}{n_i^2}\right) = \frac{R}{n_f^2} \Rightarrow \lambda = \frac{n_f^2}{R} = \frac{4}{1.097\times 10^{7}\text{ m}^{-1}} = 3.65\times 10^{-7}\text{ m} = 365\text{ nm},$$

which is in the ultraviolet. Therefore, there are $\boxed{\text{an infinite number of Balmer lines in the ultraviolet}}$. All

lines from $n_i = 10$ to ∞ all fall in the ultraviolet part of the spectrum.

28.25 Setting Equations 17.2 and 8.7 equal gives:

$$\frac{kZq_e^2}{r_n^2} = \frac{m_e v^2}{r_n},$$

so that

$$r_n = \frac{kZq_e^2}{m_e v^2} = \frac{kZq_e^2}{m_e}\frac{1}{v^2}.$$

From Equation 28.7, $m_e v r_n = n\dfrac{h}{2\pi}$, we can substitute for the velocity giving:

$$r_n = \frac{kZq_e^2}{m_e} \cdot \frac{4\pi^2 m_e^2 r_n^2}{n^2 h^2}$$

so that

$$r_n = \frac{n^2}{Z}\frac{h^2}{4\pi^2 m_e k q_e^2} = \frac{n^2}{Z}a_B, \text{ where } a_B = \frac{h^2}{4\pi^2 m_e k q_e^2},$$

which are Equations 28.8 and 28.9.

28.31 (a) Using Equation 28.15, we can calculate the angular momentum of an $\ell = 1$ electron:

$$L = \sqrt{\ell(\ell+1)}\frac{h}{2\pi} = \sqrt{1(2)}\left(\frac{6.626\times10^{-34}\text{ J}\cdot\text{s}}{2\pi}\right) = \boxed{1.49\times10^{-34}\text{ J}\cdot\text{s}}$$

(b) Using Equation 28.17, we can determine the electron's spin angular momentum, since $s = 1/2$:

$$S = \sqrt{s(s+1)}\frac{h}{2\pi} = \sqrt{\frac{1}{2}\left(\frac{3}{2}\right)}\frac{6.626\times10^{-34}\text{ J}\cdot\text{s}}{2\pi} = \boxed{9.13\times10^{-35}\text{ J}\cdot\text{s}}$$

(c) To calculate the ratio of L to S, use Equations 28.15 and 28.17:

$$\frac{L}{S} = \frac{\sqrt{\ell(\ell+1)}h/2\pi}{\sqrt{s(s+1)}h/2\pi} = \frac{\sqrt{2}}{\sqrt{3/4}} = \boxed{1.63}$$

28.37 (a) We know from Equation 28.19 that the $\ell = 1$ subshell can have $2(2\ell+1) = 2(2+1) = 6$ electrons. The $\ell = 2$ subshell can have $2(4+1) = 10$ electrons. So, $\boxed{\ell = 2}$ will be the minimum value of ℓ to have 9 electrons in it.

(b) Using the spectroscopic notation discussed on page 761, where $n = 3$, $\ell = 2$, and the number of electrons is 9, we have: $\boxed{3d^9}$.

28.43 Using Equations 18.3, 27.4 and 23.3, gives: $E = qV = \dfrac{hc}{\lambda}$, which allows us to calculate the wavelength:

$$\lambda = \frac{hc}{qV} = \frac{(6.626\times10^{-34}\text{ J}\cdot\text{s})(2.998\times10^{8}\text{ m/s})}{(1.602\times10^{-19}\text{ C})(30.0\times10^{3}\text{ V})} = \boxed{4.13\times10^{-11}\text{ m}}$$

28.49 (a) Equation 28.10 gives the energy for hydrogen-like atoms to be $E_n = -\dfrac{Z^2}{n^2}E_0$, where $E_0 = 13.6\text{ eV}$. Now, the K_α x-ray energy is given by:

$$E_{K_\alpha} = E_2 - E_1 = -\left(\frac{Z^2}{4}\right)(13.6\text{ eV}) - \left[-\left(\frac{Z^2}{1}\right)(13.6\text{ eV})\right] = Z^2(13.6\text{ eV})\left(\frac{3}{4}\right).$$

Since we know the energy, we can determine the effective Z, and therefore determine the element:

$$Z = \sqrt{\frac{4}{3}\frac{E_{K_\alpha}}{13.6\text{ eV}}} = \sqrt{\frac{4}{3}\left(\frac{52.9\times10^{3}\text{ eV}}{13.6\text{ eV}}\right)} = 72\,.$$

Therefore, the element we want would have an atomic number of 73, which is $\boxed{\text{Tantalum (Ta)}}$.

(b) If the x-ray tube had a 50 kV applied voltage, the most kinetic energy an electron can have is 50 keV. The highest-energy x ray produced is one for which all of the electron's energy is converted to photon energy, producing a 50 keV photon, which is not as energetic as the x-rays discussed in part (a).

28.55 (a) From Figure 28.49, we see that it would take 2.3 eV photons to pump chromium atoms into its second excited state. Similarly, it would take 3.0 eV photons to pump chromium atoms into its third excited state.

(b) $\lambda_2 = \dfrac{hc}{E} = \dfrac{1.24 \times 10^{-6} \text{ eV} \cdot \text{m}}{2.3 \text{ eV}} = 5.39 \times 10^{-7} \text{ m} = \boxed{5.4 \times 10^2 \text{ nm}}$, which is yellow-green.

$\lambda_3 = \dfrac{hc}{E} = \dfrac{1.24 \times 10^{-6} \text{ eV} \cdot \text{m}}{3.0 \text{ eV}} = 4.13 \times 10^{-7} \text{ m} = \boxed{4.1 \times 10^2 \text{ nm}}$, which is blue-violet.

28.61 We will use Equation 26.4 to determine the speed of the star, since we are given the observed wavelength of $\lambda_{obs} = 91.2 \text{ nm}$. We first need to calculate the source wavelength by using Equations 25.1, 27.3 and 28.10 to get:

$$\Delta E = E_f - E_i = \frac{hc}{\lambda} = \left(-\frac{Z^2}{n_f^2} E_0 \right) - \left(-\frac{Z^2}{n_i^2} E_0 \right) = 0 - \left[-\frac{1^2}{1^2} (13.6 \text{ eV}) \right] = 13.6 \text{ eV},$$

so that

$$\lambda_s = \frac{hc}{\Delta E} = \frac{1.24 \times 10^3 \text{ eV} \cdot \text{nm}}{13.6 \text{ eV}} = 91.2 \text{ nm}.$$

Therefore, using Equation 26.4, $\lambda_{obs} = \lambda_s \sqrt{\dfrac{1+u/c}{1-u/c}}$, we have $\dfrac{1+u/c}{1-u/c} = \dfrac{\lambda_{obs}^2}{\lambda_s^2}$, so that $1 + \dfrac{u}{c} = \dfrac{\lambda_{obs}^2}{\lambda_s^2}\left(1 - \dfrac{u}{c} \right)$

and thus,

$$\frac{u}{c} = \frac{\lambda_{obs}^2/\lambda_s^2 - 1}{\lambda_{obs}^2/\lambda_s^2 + 1} = \frac{(91.0 \text{ nm}/91.2 \text{ nm})^2 - 1}{(91.0 \text{ nm}/91.2 \text{ nm})^2 + 1} = -2.195 \times 10^{-3}.$$

So,

$$u = \left(-2.195 \times 10^{-3} \right)\left(2.998 \times 10^8 \text{ m/s} \right) = -6.58 \times 10^5 \text{ m/s}.$$

Since u is negative, the star is moving toward the earth at a speed of $\boxed{6.58 \times 10^5 \text{ m/s}}$.

28.67 From the definition of velocity, Equation 2.3, we can get an expression for the velocity in terms of the period of rotation of the moon: $v = \dfrac{2\pi R}{T}$. Then, from Equation 9.11 for a point object we get the angular momentum:

$$L = I\omega = mR^2\omega = mR^2 \frac{v}{R} = mRv.$$

Substituting for the velocity and setting equal to Equation 28.15 gives:

$$L = mvR = \frac{2\pi mR^2}{T} = \sqrt{\ell(\ell+1)}\frac{h}{2\pi}.$$

Since ℓ is large:

$$\frac{2\pi mR^2}{T} \approx \frac{\ell h}{2\pi} \text{ therefore, } \ell = \frac{4\pi^2 mR^2}{Th}.$$

Using the constants from the front cover of the textbook: $R = 3.84 \times 10^8 \text{ m}$, $T = 2.36 \times 10^6 \text{ s}$, and $m = 7.35 \times 10^{22} \text{ kg}$ gives:

$$\ell = \frac{4\pi^2 \left(7.35 \times 10^{22} \text{ kg} \right)\left(3.84 \times 10^8 \text{ m} \right)^2}{\left(2.36 \times 10^6 \text{ s} \right)\left(6.63 \times 10^{-34} \text{ J} \cdot \text{s} \right)} = \boxed{2.73 \times 10^{68}}.$$

28.73 We will use Equation 26.4 to determine the observed wavelengths for the Doppler shifted hydrogen line. First, for the hydrogen moving away from us, we use $u = +50.0$ km/s, so that:

$$\lambda_{obs} = (91.20 \text{ nm}) \sqrt{\frac{1 + \left(50.0 \times 10^3 \text{ m/s} / 2.998 \times 10^8 \text{ m/s}\right)}{1 - \left(50.0 \times 10^3 \text{ m/s} / 2.998 \times 10^8 \text{ m/s}\right)}} = 91.22 \text{ nm}.$$

Then, for the hydrogen moving toward us, we use $u = -50.0$ km/s, so that:

$$\lambda_{obs} = (91.20 \text{ nm}) \sqrt{\frac{1 - \left(50.0 \times 10^3 \text{ m/s} / 2.998 \times 10^8 \text{ m/s}\right)}{1 + \left(50.0 \times 10^3 \text{ m/s} / 2.998 \times 10^8 \text{ m/s}\right)}} = 91.18 \text{ nm}.$$

So the range of wavelengths is from $\boxed{91.18 \text{ nm to } 91.22 \text{ nm}}$.

RADIOACTIVITY AND NUCLEAR PHYSICS

29

CONCEPTUAL QUESTIONS

29.1 Radioactivity is found to be associated with certain elements, such as uranium. Radiation does not vary with chemical state – that is, uranium is radioactive whether it is an element or in a compound. Additionally, radiation does not vary with temperature, pressure, or ionization state of the uranium atom. The huge energy emitted in each event is another piece of evidence that radioactivity cannot be atomic.

29.4 In air, the α particles do not interact with very much, and can travel a few centimeters. β particles in lead travel about a millimeter because of their interaction with the dense material. So, β particles travel a shorter distance in lead than α particles travel in air because the β particles collide more in the lead and therefore transfer more energy to the charged material through collisions.

29.7 γ radiation comes from the deexcitation of a nucleus, just as an x-ray comes from the deexcitation of an atom. The names γ rays and x-rays identify the source of the radiation. At the same energy, γ rays and x-rays are otherwise identical. γ rays can be more energetic than x-rays because they come from nuclear deexcitations rather than atomic ones. There is a range over which both γ rays and x-rays can exist. (See Figure 23.7)

29.10 Both protons and neutrons exist in the nucleus and have the same intrinsic spin. The proton, however, is positively charged, while the neutron is neutral. Also, even though their masses are approximately the same, the neutron is slightly more massive once you look past the third digit.

29.13 Nuclei having the same Z and different Ns are defined to be isotopes of the same element. Since they are the same element, they have the same number of protons, and therefore the same number of electrons when neutral. Thus, different isotopes of the same element will have similar chemistries because their atomic properties are the same.

29.16 This reaction is an electron capture, so Equation 29.8 tells us that the neutrino produced in the reaction will be a ν_e, or an electron's neutrino. Since an electron is captured, it is necessary to produce an electron neutrino in order to conserve electron family number. Since the proton is converted into a neutron, the total number of nucleons is conserved, as is expected.

29.19 The radioactive isotopes: ^{226}Ra, ^{222}Rn, and ^{210}Po are found in the rock because the decay chain of ^{238}U constantly produces them. Figure 29.14 tells how these particular isotopes are formed. They are still present even though they decay much faster than the uranium because each time ^{238}U decays it eventually produces these isotopes.

29.22 One kilogram of uranium contains more atoms of uranium than one kilogram of uranium hexafluoride because there are no additional atoms in the mass. Therefore, one kilogram of uranium should be more radioactive than one kilogram of uranium hexafluoride, but one mole of uranium will not be more radioactive than one mole of uranium hexafluoride.

29.25 Taking into account the binding energy of the electrons in the neutral atoms would increase the binding energy of the nuclide. To separate a hydrogen atom into a proton and an electron requires 13.6 eV of energy. Therefore, the first term in Equation 29.17 would be larger since the electrons would need to be separated from the protons. This difference is equal to Z times 13.6 eV, which is at least four orders of magnitude smaller than the binding energies of the nucleons, so it is negligible.

29.28 No, the α particle tunnels through the barrier rather than traveling each point along an imaginary line from inside to out. An α particle outside the range of the nuclear force feels the repulsive Coulomb force. The α does not have enough kinetic energy to get over the rim of the potential barrier, yet it does manage to get out by quantum mechanical tunneling.

PROBLEMS

29.1 To calculate the number of pairs created, simply divide the total energy by the energy needed per pair:

$$\text{\# ion pairs} = \frac{(0.500 \text{ MeV})(1.00\times10^6 \text{ eV/MeV})}{30.0 \text{ eV/pair}} = \boxed{1.67\times10^4 \text{ pairs}}.$$

This is the maximum number of ion pairs because it assumes that all the energy goes to creating ion pairs and that there are no energy losses.

29.7 (a) Using Equation 29.3, we can approximate the radius of Ni-58:

$$r_{Ni} = r_0 A_{Ni}^{1/3} = \left(1.2\times10^{-15} \text{ m}\right)(58)^{1/3} = 4.6\times10^{-15} \text{ m} = \boxed{4.6 \text{ fm}}$$

(b) Again, using Equation 29.3, this time we can approximate the radius of Ha-258:

$$r_{Ha} = \left(1.2\times10^{-15} \text{ m}\right)(258)^{1/3} = 7.6\times10^{-15} \text{ m} = 7.6 \text{ fm}.$$

Finally, taking the ratio of Ni to Ha gives:

$$\frac{r_{Ni}}{r_{Ha}} = \frac{4.645\times10^{-15} \text{ m}}{7.639\times10^{-15} \text{ m}} = \boxed{0.61}.$$

29.13 (a) Recalling the definition of density from Equation 10.1 and recalling the formula for the volume of a sphere,

$$\rho = \frac{M}{V} = \frac{M}{(4/3)\pi r^3},$$ we can calculate the radius of the neutron star:

$$r = \left[\frac{(3/4)M}{\pi\rho}\right]^{1/3} = \left[\frac{(0.750)(1.3)(1.99\times10^{30} \text{ kg})}{\pi\left(2.3\times10^{17} \text{ kg/m}^3\right)}\right]^{1/3} = 1.39\times10^4 \text{ m} = \boxed{14 \text{ km}}$$

(b) Using Equation 29.3, $r = r_0 A^{1/3}$, we can solve for the atomic mass, using the radius calculated in part (a):

$$A = \left(\frac{r}{r_0}\right)^3 = \left(\frac{1.39\times10^4 \text{ m}}{1.2\times10^{-15} \text{ m}}\right)^3 = \boxed{1.6\times10^{57}}.$$

29.19 Referring to Equation 29.8, we need to calculate the values for Z and N. From the periodic table, we know that Indium has $Z = 49$, and the element with $Z = 48$ is Cadmium (Cd). Using Equation 29.2, we know that $N = A - Z = 106 - 49 = 57$ for In, and $N = 58$ for Cd. Putting this all together gives:

$$\boxed{{}^{106}_{49}\text{In}_{57} + e^- \rightarrow {}^{106}_{48}\text{Cd}_{58} + \nu_e}$$

29.25 Since we know that Pb-208 is the product of an alpha decay, Equation 29.4 tells us that: $A - 4 = 208$ and since $Z - 2 = 82$ from the periodic table, we then know that $N - 2 = 208 - 82 = 126$. So, for the parent nucleus we have: $A = 212$, $Z = 84$, and $N = 128$. Therefore, from the periodic table, the parent nucleus is $\boxed{\text{Polonium-212}}$ and the decay equation is:

$$\boxed{{}^{212}_{84}\text{Po}_{128} \rightarrow {}^{208}_{82}\text{Pb}_{126} + {}^{4}_{2}\text{He}_2}.$$

29.31 (a) First, we know from the periodic table that $Z = 88$ for radium. Then, since the decay equation is:

$$^{222}_{88}\text{Ra} \rightarrow {}^{A}_{Z}\text{X}_N + {}^{14}_{6}\text{C}_8,$$

we know that: $A = 222 - 14 = 208$; $Z = 88 - 6 = 82$; and $N = A - Z = 208 - 82 = 126$, so from the periodic table, the element is lead, and $X = \boxed{{}^{208}_{82}\text{Pb}_{126}}$.

(b) The energy emitted in the decay is found by determining the change in mass, then using Equation 29.5. First, we can get the masses from Appendix A:

$$\Delta m = m\left({}^{222}_{88}\text{Ra}_{134}\right) - m\left({}^{208}_{82}\text{Pb}_{126}\right) - m\left({}^{14}_{6}\text{C}_8\right)$$

$$= 222.015353\text{u} - 207.976627\text{u} - 14.003241\text{u} = 3.5485 \times 10^{-2}\text{u}$$

Equation 29.5 then gives the energy emitted:

$$E = \Delta mc^2 = \left(3.5485 \times 10^{-2}\text{ u}\right)\left(\frac{931.5\text{ MeV}/c^2}{\text{u}}\right)c^2 = \boxed{33.05\text{ MeV}}.$$

29.37 (a) Using Equation 29.7 and the periodic table, we can get the complete decay equation:

$$^{11}_{6}\text{C}_5 \rightarrow {}^{11}_{5}\text{B}_6 + \beta^+ + \nu_e.$$

(b) To calculate the energy emitted we first need to calculate the change in mass. The change in mass is the mass of the parent minus the mass of the daughter and the positron it created. The masses given for the parent and daughter, however, are given for neutral atoms, so the carbon has one additional electron than the boron and we must subtract an additional mass of the electron to get the correct change in mass:

$$\Delta m = m\left({}^{11}\text{C}\right) - \left[m\left({}^{11}\text{B}\right) + 2m_e\right] = 11.011433\text{ u} - \left[11.009305\text{ u} + 2\left(0.00054858\text{ u}\right)\right] = 1.031 \times 10^{-3}\text{ u}.$$

Then, Equation 29.5 gives the energy emitted:

$$E = \Delta mc^2 = \left(1.031 \times 10^{-3}\text{ u}\right)\left(\frac{931.5\text{ MeV/u}}{c^2}\right)c^2 = \boxed{0.9602\text{ MeV}}$$

29.43 (a) First, we must determine the number of atoms we have of Radium. We use the molar mass of 226 g/mol to

get: $N = (1.00 \text{ g})\left(\dfrac{\text{mol}}{226 \text{ g}}\right)\dfrac{6.022\times10^{23} \text{ atoms}}{\text{mol}} = 2.6646\times10^{21}$ atoms . Then, using Equation 29.15,

$R = \dfrac{0.693N}{t_{1/2}}$, where we know the half-life of Radium-226 is 1.60×10^{3} y , from Appendix B:

$$R = \dfrac{(0.693)(2.6646\times10^{21})}{1.60\times10^{3} \text{ y}} \times \left(\dfrac{1 \text{ y}}{3.156\times10^{7} \text{ s}}\right) = 3.66\times10^{10} \text{ Bq}\left(\dfrac{\text{Ci}}{3.70\times10^{10} \text{ Bq}}\right) = \boxed{0.988 \text{ Ci}}$$

(b) The half-life of ^{226}Ra is now known more accurately than it was when the Ci unit was established.

29.49 Using Equation 29.15, we can write the activity in terms of the half-life, the molar mass, M, and the mass of the sample, m:

$$R = \dfrac{0.693N}{t_{1/2}} = \dfrac{(0.693)\left[\left(6.02\times10^{23} \text{ atoms/mol}\right)/M\right]m}{t_{1/2}} .$$

From the periodic table, we know that $M = 50.94$ g/mol , so we can determine the half-life:

$$t_{1/2} = \dfrac{(0.693)\left(6.022\times10^{23} \text{ atoms/mol}\right)(1000 \text{ g})}{(50.94 \text{ g/mol})(1.75 \text{ Bq})} = 4.681\times10^{24} \text{ s}\left(\dfrac{1 \text{ y}}{3.156\times10^{7} \text{ s}}\right) = \boxed{1.48\times10^{17} \text{ y}}$$

29.55 (a) Using Equation 29.15, we can write the activity in terms of the half-life, the atomic mass, M, and the mass of the sample, m: $R = \dfrac{0.693N}{t_{1/2}} = \dfrac{(0.693)(m/M)}{t_{1/2}}$. The atomic mass of tritium (from Appendix A) is:

$$M = 3.016050 \text{ u}\left(\dfrac{1.6605\times10^{-27} \text{ kg}}{1 \text{ u}}\right) = 5.0082\times10^{-27} \text{ kg/atom} ,$$

and the half-life is 12.33 y (from Appendix B), so we can determine the original mass of tritium:

$m = \dfrac{Rt_{1/2}M}{0.693}$ or

$$m = \dfrac{(15.0 \text{ Ci})(12.33 \text{ y})\left(5.0082\times10^{-27} \text{ kg}\right)}{0.693}\left(\dfrac{3.70\times10^{10} \text{ Bq}}{\text{Ci}}\right)\left(\dfrac{3.156\times10^{7} \text{ s}}{\text{y}}\right) = 1.56\times10^{-6} \text{ kg} = \boxed{1.56 \text{ mg}}$$

(b) Using Equations 29.16 and 29.11, we can calculate the activity after 5.00 years:

$$R = R_0 e^{-\lambda t} = R_0 \exp\left(-\dfrac{0.693}{t_{1/2}}t\right) = (15.0 \text{ Ci})\exp\left(\dfrac{-0.693(5.00 \text{ y})}{12.33 \text{ y}}\right) = \boxed{11.3 \text{ Ci}}$$

29.61 Dividing Equation 29.17 by A gives the binding energy per nucleon:

$$\dfrac{\text{BE}}{A} = \dfrac{\left[Zm\left(^{1}\text{H}\right) + Nm_n - m\left(^{209}_{83}\text{Bi}_{126}\right)\right]c^2}{A} .$$

We know that $Z = 83$ (from the periodic table), $N = A - Z = 209 - 83 = 126$, and the atomic mass of the Bi-209 nuclide is 208.980374 u (from Appendix A), so that:

$$\dfrac{\text{BE}}{A} = \dfrac{\left[83(1.007825 \text{ u}) + 126(1.008665 \text{ u}) - 208.980374 \text{ u}\right]}{209}\left(\dfrac{931.5 \text{ MeV/u}}{c^2}\right)c^2 = \boxed{7.848 \text{ MeV/nucleon}}$$

This binding energy per nucleon is approximately the value given in the graph.

29.67 Using Equation 21.3, we can determine the radius of a moving charge in a magnetic field: $r = \dfrac{mv}{qB}$. First, we need to determine the velocity of the proton. Since the energy of the proton (10.0 MeV) is substantially less than the rest mass energy of the proton (938 MeV), we know the velocity is non-relativistic and that $E = \dfrac{1}{2}mv^2$.

Therefore,

$$v = \left(\frac{2E}{m}\right)^{1/2} = \left(\frac{2 \cdot 10.0 \text{ MeV}}{938.27 \text{ MeV}/c^2}\right)^{1/2} = (0.1460)(2.998 \times 10^8 \text{ m/s}) = 4.377 \times 10^7 \text{ m/s}.$$

So,

$$r = \frac{\left(1.6726 \times 10^{-27} \text{ kg}\right)\left(4.377 \times 10^7 \text{ m/s}\right)}{\left(1.602 \times 10^{-19} \text{ C}\right)(2.00 \text{ T})} = 0.228 \text{ m} = \boxed{22.8 \text{ cm}}$$

29.73 We know that the kinetic energy for a relativistic particle is given by Equation 26.7, $KE_{rel} = (\gamma - 1)mc^2$, and that

since $\gamma = \left(1 - \dfrac{v^2}{c^2}\right)^{-1/2}$, we can get an expression for the speed:

$$\frac{v^2}{c^2} = \left(1 - \frac{1}{\gamma^2}\right), \text{ or } v = c\sqrt{1 - \frac{1}{\gamma^2}}.$$

For β particle: $KE = 5.00 \text{ MeV} = (\gamma - 1)(0.511 \text{ MeV})$, so that $(\gamma - 1) = 9.785$ or $\gamma = 10.785$. Thus, the velocity for the β particle is:

$$v_\beta = c\sqrt{1 - \frac{1}{\gamma^2}} = \left(2.998 \times 10^8 \text{ m/s}\right)\left(1 - \frac{1}{(10.785)^2}\right)^{1/2} = 2.985 \times 10^8 \text{ m/s}.$$

For α particle: $KE = 5.00 \text{ MeV} = (\gamma - 1)(4.0026 \text{ u})\left(\dfrac{931.5 \text{ MeV/u}}{c^2}\right)c^2$ so that $(\gamma - 1) = 1.341 \times 10^{-3}$ or

$\gamma = 1.00134$. Thus, the velocity for the α particle is:

$$v_\alpha = \left(2.998 \times 10^8 \text{ m/s}\right)\left(1 - \frac{1}{(1.00134)^2}\right)^{1/2} = 1.551 \times 10^7 \text{ m/s}.$$

Finally, the ratio of the velocities is given by:

$$\frac{v_\beta}{v_\alpha} = \frac{2.985 \times 10^8 \text{ m/s}}{1.551 \times 10^7 \text{ m/s}} = \boxed{19.2}.$$

In other words, when the β and α particles have the same kinetic energy, the β particle is approximately 19 times faster than the α particle.

29.79 (a) Using Equation 29.17, we can calculate the binding energy, assuming that the particle is made from two neutrons:

$$BE = \left[2m_n - m(\text{particle})\right]c^2 = \left[2(1.008665 \text{ u}) - 2.02733 \text{ u}\right]\left(\frac{931.5 \text{ MeV/u}}{c^2}\right)c^2 = \boxed{-9.315 \text{ MeV}}.$$

(b) The binding energy cannot be negative; the nucleons would not stay together.

(c) The particle cannot be made from two neutrons.

30 APPLICATIONS OF NUCLEAR PHYSICS

CONCEPTUAL QUESTIONS

30.1 It is much easier to spot one big burst of energy than several smaller ones.

30.4 For medical imaging, it is necessary to detect a certain amount of radiation, so a dye is given to the patient containing a radioactive isotope, which is then detected. If the isotope has a short half-life, then there will be more decays from a smaller amount of that isotope, and so the patient will need to ingest a smaller amount of that isotope. In other words, a smaller half-life means that the decays occur during the imaging process rather than for a substantial amount of time after the imaging is over, thus limiting the dose to the patient.

30.7 From Equations 30.3, we see that the reaction involving tritium produces more energy than the reaction involving deuterium exclusively.

30.10 Energy is released if the products of a nuclear reaction have a greater binding energy per nucleon (BE/A) than the parent nuclei. Since the BE/A is greater for medium-mass nuclei than heavy nuclei, when heavy nuclei split, the products have less mass per nucleon, so that mass is destroyed and energy is released in the reaction. If light-mass nuclei fission, the BE/A of the products is smaller than that for the parent nuclei and therefore energy is required to produce the reaction.

30.13 Radioactive iodine was one of the three greatest radioactive waste products of the Chernobyl disaster and since iodine is concentrated in the thyroid, the disaster led to an increase in thyroid cancers. Since iodine is present in milk and children tend to ingest more milk, they would be more susceptible to thyroid cancers.

30.16 Neutron flux must be carefully regulated to avoid an exponential increase in fissions, called supercriticality. Control rods help to prevent overheating. Water used to thermalize neutrons, necessary to get them to induce fission in ^{235}U and achieve criticality, provides a negative feedback for temperature increases. Should the reactor overheat and boil the water to steam or be breached, the absence of water kills the chain reaction.

30.19 A shaped charge will cause equal momentum to be given to both the large back plate and the lighter weight target of the blast. This means that the lighter weight target will carry a proportionally larger velocity and hence a proportionally larger kinetic energy since $KE = \dfrac{1}{2}mV^2$.

30.22 The environmental activities of these two isotopes are decreasing faster than their half-lives because the concentration of these two isotopes is decreasing faster than their half-lives. This might be occurring because both isotopes are heavy and they therefore have settled out of the atmosphere into the ground, rivers, etc. So, if core samples were taken of the ground, one would expect activities of these two isotopes to be higher because of this settling effect.

30.25 Because blood has higher sensitivity to radiation than many other parts of the body, the latency period for leukemia, which is a blood disease, is shorter than for most other forms of cancer.

30.28 For the x-ray fluorescence machine, the only type of radiation protection used is limiting the time of exposure. Since the machine turns off when there are no feet in them, this limits the exposure to just the time the machine is used. For the nuclear plant employee, the type of radiation protection used is shielding. The protective clothing helps to reduce his exposure to the high-doses of radiation when he must be close to it. Also, he limits his time in the high-dose area to keep his exposure to a minimum.

30.31 Abnormal tissue is more sensitive to radiation than normal tissue. Also, since some parts of the body are more sensitive to radiation than other parts, it would seem reasonable that some cancers, involving those parts of the body, would be more sensitive to radiation as well.

30.34 Cobalt-60 γ rays average 1.25 MeV, while those of ^{137}Cs are 0.660 MeV and are less penetrating. Since the ^{137}Cs produces a smaller amount of energy per γ, more time is needed to get the same dose.

PROBLEMS

30.1 Using Equation 26.6, we can determine the energy output of the reaction by calculating the change in mass of the constituents in the reaction, where the masses are found either in Appendix A or Table 29.1:

$$E = \left(m_i - m_f\right)c^2 = \left(4.002603 \text{ u} + 9.012182 \text{ u} - 12.000000 \text{ u} - 1.008665 \text{ u}\right)\left(931.5 \text{ MeV/u}\right) = \boxed{5.701 \text{ MeV}}$$

30.7 Beginning with Equation 29.15,

$$R = \frac{0.693N}{t_{1/2}} = \frac{(0.693)(m/M)N_A}{t_{1/2}},$$

we can solve for the mass of the iodine isotopes, where the atomic masses and the half lives are given in the appendices:

$$m_{131} = \frac{RM\,t_{1/2}}{0.693N_A} = \frac{\left(50\times10^{-6}\text{ Ci}\right)\left(3.70\times10^{10}\text{ Bq/Ci}\right)(130.91\text{ g/mol})(8.04\text{ d})\left(8.64\times10^4\text{ s/d}\right)}{(0.693)\left(6.02\times10^{23}\text{ /mol}\right)} = \boxed{4.0\times10^{-10}\text{ g}}$$

$$m_{123} = \frac{RM\,t_{1/2}}{0.693N_A} = \frac{\left(70\times10^{-6}\text{ Ci}\right)\left(3.70\times10^{10}\text{ Bq/Ci}\right)(122.91\text{ g/mol})(13.2\text{ h})(3600\text{ s/h})}{(0.693)\left(6.02\times10^{23}\text{ /mol}\right)} = \boxed{3.6\times10^{-11}\text{ g}}$$

30.13 From Table 6.1, we know the annual U.S. energy use is $E = 8 \times 10^{19}$ J and from Example 30.1, we know that a 1.00 kg mixture of deuterium and tritium releases 3.37×10^{14} J of energy, so we can determine the mass of deuterium and tritium required to supply the annual energy for the U.S.:

$$M = \left(8 \times 10^{19} \text{ J}\right)\left(\frac{1.00 \text{ kg}}{3.37 \times 10^{14} \text{ J}}\right) = \boxed{2 \times 10^{5} \text{ kg}} \quad 200 \text{ tons}.$$

30.19 (a) The number of protons consumed per second can be determined by examining Equations 30.1, and realizing that 4 protons are needed for each cycle to occur. The energy released by a proton-proton cycle is given in Equation 30.2, so that:

$$\text{\# protons/s} = (0.90)\left(4 \times 10^{26} \text{ J/s}\right)\left(\frac{4 \text{ protons}}{26.7 \text{ MeV}}\right)\left(\frac{1 \text{ MeV}}{1.602 \times 10^{-13} \text{ J}}\right) = \boxed{3 \times 10^{38} \text{ protons/s}}$$

(b) Examining Equation 30.2, tells us that for each cycle, 2 neutrinos are created and 4 protons are destroyed. To determine the number of neutrinos at the surface of the earth, we need to determine the number of neutrinos leaving the sun and divide that by the surface area of a sphere with radius the distance from the sun to the earth:

$$S = \frac{\#}{\text{area}} = \frac{\#}{4\pi R^2} = \frac{\left(2 v_e / 4 \text{ protons}\right)\left(3.37 \times 10^{38} \text{ protons/s}\right)}{4\pi \left(1.50 \times 10^{11} \text{ m}\right)^2} = \boxed{\frac{\left(6 \times 10^{14}\right) v_e / \text{m}^2}{\text{s}}}$$

30.25 (a) To calculate the energy released, we use Equation 26.6 to calculate the difference in energy before and after the reaction:

$$E = \left(m_n + m\left(^{239}\text{Pu}\right) - m\left(^{96}\text{Sr}\right) - m\left(^{140}\text{Ba}\right) - 4m_n\right)c^2 = \left(m\left(^{239}\text{Pu}\right) - m\left(^{96}\text{Sr}\right) - m\left(^{140}\text{Ba}\right) - 3m_n\right)c^2.$$

Appendix A and Table 29.1 will give the masses not provided in the problem statement:

$$E = \left[239.052157 \text{ u} - 95.921750 \text{ u} - 139.910581 \text{ u} - (3)(1.008665 \text{ u})\right](931.5 \text{ MeV/u}) = \boxed{180.6 \text{ MeV}}$$

(b) Writing the equation in full form gives:

$$_0^1 n + {}_{94}^{239}\text{Pu}_{145} \rightarrow {}_{38}^{96}\text{Sr}_{58} + {}_{56}^{140}\text{Ba}_{84} + 4\,{}_0^1 n,$$

so we can determine the total number of nucleons before and after the reaction:

$$A_i = 1 + 239 = 240 = 96 + 140 + 4 = A_f,$$

and the total charge before and after the reaction:

$$Z_i = 0 + 94 = 94 = 56 + 38 + 4(0) = Z_f.$$

Therefore, both the total number of nucleons and the total charge are conserved.

30.31 Using Equation 26.6, we can calculate the mass converted into energy for a 12.0 kT bomb:

$$m = \frac{E}{c^2} = \frac{(12.0 \text{ kT})\left(4.2 \times 10^{12} \text{ J/kT}\right)}{\left(2.998 \times 10^8 \text{ m/s}\right)^2} = 5.61 \times 10^{-4} \text{ kg} = \boxed{0.56 \text{ g}}$$

30.37 (a) For fission reactions, the energy produced is 200 MeV per fission, we can convert the ¼ of 300 kT yield into number of fissions:

$$\text{\# of fissions} = \frac{(1/4)(300 \text{ kT})(4.2 \times 10^{12} \text{ J/kT})}{(200 \text{ MeV/fission})(1.602 \times 10^{-13} \text{ J/MeV})} = \boxed{9.8 \times 10^{24} \text{ fissions}} .$$

Then, since we know the average molar mass, we can determine the mass in kilograms:

$$m = (9.83 \times 10^{24} \text{ nuclei}) \left(\frac{1 \text{ mol}}{6.022 \times 10^{23} \text{ nuclei}} \right) (238 \text{ g/mol}) = 3.89 \times 10^{3} \text{ g} = \boxed{3.9 \text{ kg}}$$

(b) Similarly, given that for fusion reactions, the energy produced is 20 MeV per fusion, we can convert the ¾ of 300 kT yield into number of fusions:

$$\text{\# of fusions} = \frac{(3/4)(300 \text{ kT})(4.2 \times 10^{12} \text{ J/kT})}{(20 \text{ MeV/fusion})(1.602 \times 10^{-13} \text{ J/MeV})} = 2.95 \times 10^{26} \text{ fusions} = \boxed{2.9 \times 10^{26} \text{ fusions}} .$$

Then, since we know the average molar mass of the two nuclei in each reaction, we can determine the mass:

$$m = (2.95 \times 10^{26} \text{ fusions}) \left(\frac{1 \text{ mol}}{6.022 \times 10^{23} \text{ nuclei}} \right) (5 \text{ g LiD fuel/mol}) = 2.45 \times 10^{3} \text{ g} = \boxed{2 \text{ kg}}$$

(c) The nuclear fuel totals only 6 kg, so it is quite reasonable that some missiles carry 10 warheads. The mass of the fuel would only be 60 kg and therefore the mass of the 10 warheads, weighing say 10 times the nuclear fuel, would be only 1500 lbs. If the fuel for the missiles weighs, say 5 times the total weight of the warheads, the missile would weight about 9000 lbs or 4.5 tons. This is not an unreasonable weight for a missile.

30.43 Using Equation 30.6, $\text{rem} = \text{rad} \times \text{RBE}$, and from Table 30.2, we know that $\text{RBE} = 20$ for whole body exposure, so:

$$\text{rad} = \frac{\text{rem}}{\text{RBE}} = \frac{4000 \text{ rem}}{20} = \boxed{2.0 \times 10^{2} \text{ rad}} .$$

30.49 *Step 1:* This problem involves a person ingesting Ra-226, and that produces α radiation.

Step 2: Calculate the yearly dose in rem to the bones.

Step 3: Given 1.00 mg sample of Ra-226. From Appendix A, we know the atomic mass of Ra-226 is 226.025402 u, and its half-life is 1.6×10^{3} y. Also, we know that there is 8.00 kg of bone tissue.

Step 4: To determine the energy deposited per year, we first need to determine the activity. Using Equation 29.15, we have:

$$R = \frac{0.693N}{t_{1/2}} = \frac{(0.693)(m/M)}{t_{1/2}} = \frac{(0.693)(1.00 \times 10^{-6} \text{ kg}/226.025402 \text{ u})}{(1.6 \times 10^{3} \text{ y})(3.156 \times 10^{7} \text{ s/y})} \left(\frac{1 \text{ u}}{1.6605 \times 10^{-27} \text{ kg}} \right) = 3.66 \times 10^{7} \text{ Bq} .$$

Next, since we know that each decay produces 4.80 MeV of energy, we can calculate the total energy deposited per year:

$$E/\text{y} = (3.657 \times 10^{7} \text{ Bq}) \left(\frac{4.80 \text{ MeV}}{\text{decay}} \right) \left(\frac{3.156 \times 10^{7} \text{ s}}{\text{y}} \right) = 5.54 \times 10^{15} \text{ MeV/y} \left(\frac{1.602 \times 10^{-13} \text{ J}}{\text{MeV}} \right) = 887 \text{ J/y} .$$

Step 5: Dividing the energy by the mass of tissue gives:

$$\text{dose in rad/y} = \frac{887 \text{ J/y}}{8.00 \text{ kg}} \left(\frac{1 \text{ rad}}{0.0100 \text{ J/kg}} \right) = 1.11 \times 10^{4} \text{ rad/y} .$$

Step 6: Using Table 30.2, we know that the $\text{RBE} = 20$ for α radiation, we can calculate the dose in rem:

$$\text{rem/y} = \text{rad/y} \times \text{RBE} = (1.11 \times 10^{4} \text{ rad/y}) \times (20) = \boxed{2.2 \times 10^{5} \text{ rem/y}} .$$

Step 7: This is equivalent to 600 rem/day, which is a deadly dose of radiation, so it is not surprising that these cisterns are not used any more!!!

30.55 First, we need to determine the number of decays per day:

$$\text{decays/day} = \left(50.0 \times 10^{-6}\ \text{Ci}\right)\left(3.70 \times 10^{10}\ \text{Bq/Ci}\right)\left(8.64 \times 10^{4}\ \text{s/d}\right) = 1.598 \times 10^{11}/\text{d}.$$

Next, we can calculate the energy because each decay emits an average of 0.550 MeV of energy:

$$E/\text{day} = \left(\frac{1.598 \times 10^{11}\ \text{decays}}{\text{d}}\right)(0.400)\left(\frac{0.550\ \text{MeV}}{\text{decay}}\right)\left(\frac{1.602 \times 10^{-13}\ \text{J}}{\text{MeV}}\right) = 5.633 \times 10^{-3}\ \text{J/d}.$$

Then, dividing by the mass of tissue gives the dose:

$$\text{dose in rad/d} = \frac{\left(5.633 \times 10^{-3}\ \text{J/d}\right)}{(75.0\ \text{kg})}\frac{1\ \text{rad}}{0.0100\ \text{J/kg}} = 7.51 \times 10^{-3}\ \text{rad/d}.$$

Finally, from Table 30.2, we see that the RBE = 1 for γ radiation, so:

$$\text{rem/d} = \text{rad} \times \text{RBE} = \left(7.51 \times 10^{-3}\ \text{rad/d}\right) \times (1) = 7.51 \times 10^{-3}\ \text{rem/d} = \boxed{7.5\ \text{mrem/d}}.$$

This dose is approximately 2700 mrem/y, which is larger than background radiation sources, but smaller than doses given for cancer treatments.

30.61 Recall from Equation 15.13, $I = \dfrac{P}{A}$, and from Equation 6.9, $P = \dfrac{W}{t}$, so that:

$$E = P\Delta t = IA\Delta t.$$

Therefore, we can calculate how many photons by dividing the total energy by the energy per photon:

$$\# \text{ of photons} = \frac{E}{E_{\text{photon}}} = \frac{IA\Delta t}{E_{\text{photon}}} = \frac{\left(1.50\ \text{W/m}^2\right)\left(0.0750\ \text{m}^2\right)(0.250\ \text{s})}{(100\ \text{keV/photon})\left(1.602 \times 10^{-16}\ \text{J/keV}\right)} = \boxed{1.76 \times 10^{12}\ \text{photons}}$$

30.67 (a) First, we can calculate the total energy delivered by multiplying the number of photons by their energy:

$$E = (0.750)\left(20.0 \times 10^{6}\ \gamma\text{s}\right)(0.160\ \text{MeV}/\gamma)\left(1.602 \times 10^{-13}\ \text{J/MeV}\right) = 3.84 \times 10^{-7}\ \text{J}.$$

So, dividing by the mass of affected tissue gives:

$$\text{dose in rad} = \frac{3.84 \times 10^{-7}\ \text{J}}{10.0\ \text{kg}}\frac{1\ \text{rad}}{0.0100\ \text{J/kg}} = 3.84 \times 10^{-6}\ \text{rad}.$$

Finally, from Table 30.2, we see that the RBE = 1 for γ radiation, so:

$$\text{rem} = \text{rad} \times \text{RBE} = \left(3.84 \times 10^{-6}\ \text{rad}\right) \times (1) = \boxed{3.84 \times 10^{-6}\ \text{rem}}.$$

(b) This dose is much too small. (see Table 30.4)

(c) 20,000,000 photons is not very many; this assumed number is too small.

PARTICLE PHYSICS

CONCEPTUAL QUESTIONS

31.1 The total energy in the beam is the sum of the energies of each of the individual beam particles. Since two individual particles must collide in order to produce a single particle, only the energy of the individual beam particle can be harvested, not the total energy of the beam.

31.4 The advantage of colliding beams is that having head-on collisions between particles moving in opposite directions can be designed to have total incoming momentum equal to zero so that particles with masses equivalent to twice the beam energy can be created. A collision into a fixed target means that you can have a much denser target, leading to higher chances of collision. Therefore, the disadvantage of colliding beams is that the chances of collision are smaller than fixed target collisions.

31.7 Massless particles travel at the speed of light. Time dilation approaches infinity as velocity approaches the speed of light. This implies that time does not pass for massless particles, and so they cannot self-annihilate. Thus, if neutrinos spontaneously decayed into other particles, then they must have a mass.

31.10 The antineutron should have the same lifetime as the neutron, so from Table 31.2, we expect the lifetime of an antineutron to be 887 s.

31.13 Since all nucleons are baryons, conservation of baryon number is the same thing as conservation of nucleon number which is the same as conservation of total atomic mass.

31.16 Yes, the weak nuclear force can change the flavor of a quark. If strangeness changes then strange quarks can be changed into up or downs and visa versa.

31.19 Since it is possible to separate hadrons, but it is not possible to separate quarks, the gluon force between quarks is greater than the strong nuclear force between hadrons. Since we do not observe colored objects and single quarks have color, we have been unable to separate hadrons into their constituent quark particles. In other words, quarks are confined in colorless particles.

31.22 Since the weak force is responsible for β decay, if leptons are created this implies that the weak force is acting.

31.25 Both Σ^0 and Λ^0 are composed of the quarks uds. Since the quark composition did not change and the decay gives off a photon, this implies that Σ^0 is an excited state of Λ^0.

31.28 One of the predictions of electroweak unification was the existence of the W^+, W^-, and Z^0 carrier particles. Not only were three particles having spin 1 predicted, the mass of the W^+ and W^- were predicted to be $81\,\text{GeV}/c^2$, and that of the Z^0 to be $90\,\text{GeV}/c^2$. In 1983 these carrier particles were observed at CERN with the predicted characteristics, including masses having the predicted values.

31.31 No, the existence of the Higgs boson would support the electroweak unification theory which says that the W^+, W^-, and Z^0 (the carriers of the weak force) would behave very similar to the massless photon (the carrier of the electromagnetic force) at high energies.

PROBLEMS

31.1 (a) From Table 31.1, we know that the ratio of the weak force to the electromagnetic force is:

$$\frac{\text{Weak}}{\text{Electromagnetic}} = \frac{10^{-13}}{10^{-2}} = \boxed{10^{-11}}.$$

In other words, the weak force is 11 orders of magnitude weaker than the electromagnetic force.

(b) When the forces are unified, the idea is that the four forces are just different manifestations of the same force, so under circumstances in which the forces are unified, the ratio becomes $\boxed{1 \text{ to } 1}$. (See Section 31.6)

31.7 Using the definition of velocity, we can determine the distance traveled by the W^- in a bubble chamber:

$$d = vt = (0.900)(3.00 \times 10^8 \text{ m/s})(5.00 \times 10^{-25} \text{ s}) = 1.35 \times 10^{-16} \text{ m} = \boxed{0.135 \text{ fm}}$$

31.13 If the π^0 is at rest when it decays, its initial momentum is zero, so by conservation of momentum:

$$p_i = 0 = p_f, \text{ so that } p_{\gamma_1} + p_{\gamma_2} = 0, \text{ or } p_{\gamma_1} = -p_{\gamma_2}.$$

Now, we can calculate the energy of each γ ray by using Equation 27.7, since a γ ray is a photon:

$$E_\gamma = |p_\gamma| c.$$

Therefore, since the momentums are equal in magnitude the energies of the γ rays are equal:

$$E_1 = E_2.$$

Then, by conservation of energy, the initial energy of the π^0 equals twice the energy of one of the γ rays:

$$m_{\pi^0} c^2 = 2E.$$

Finally, from Table 31.2, we can determine the rest mass energy of the π^0, and the energy of each γ ray is:

$$E = \frac{m_{\pi^0} c^2}{2} = \frac{\left(135 \text{ MeV}/c^2\right)c^2}{2} = \boxed{67.5 \text{ MeV}}.$$

31.19 (a) Using Equation 27.10, we can calculate the uncertainty in the energy, given the lifetime of the π^0 from Table 31.2:

$$\Delta E = \frac{h}{4\pi \Delta t} = \frac{6.63 \times 10^{-34} \text{ J} \cdot \text{s}}{4\pi \left(0.84 \times 10^{-16} \text{ s}\right)} = 6.28 \times 10^{-19} \text{ J} \times \frac{1 \text{ eV}}{1.60 \times 10^{-19} \text{ J}} = \boxed{3.9 \text{ eV}}$$

(b) The fraction of the decay energy is determined by dividing this uncertainty in the energy by the rest mass energy of the π^0 found in Table 31.2:

$$\frac{\Delta E}{m_{\pi^0} c^2} = \frac{3.9256 \text{ eV}}{\left(135.0 \times 10^6 \text{ eV}/c^2\right)c^2} = \boxed{2.9 \times 10^{-8}}.$$

So the uncertainty is approximately $2.9 \times 10^{-6}\%$ of the rest mass energy.

31.25 (a) From Table 31.4, we know the quark composition of each of the particles involved in this decay:

$$\Omega^-(sss) \rightarrow \Lambda^0(uds) + K^-(\bar{u}s) .$$

Then, to determine the change in strangeness, we need to subtract the initial strangeness from the final strangeness, remembering that a strange quark has a strangeness of -1:

$$\Delta S = S_f - S_i = \left[-1 + (-1)\right] - (-3) = \boxed{+1} .$$

(b) Using Table 31.3, we see that:

$$B_i = \left(\frac{1}{3} + \frac{1}{3} + \frac{1}{3}\right) = 1, \text{ and } B_f = \left(\frac{1}{3} + \frac{1}{3} + \frac{1}{3}\right) + \left(\frac{1}{3} - \frac{1}{3}\right) = 1 ,$$

so since the baryon number did not change, we know that baryon number is indeed conserved. Again, using Table 31.3, the charge is:

$$Q_i = \left(-\frac{1}{3} - \frac{1}{3} - \frac{1}{3}\right)q_e = -q_e , \text{ and } Q_f = \left(+\frac{2}{3} - \frac{1}{3} - \frac{1}{3}\right)q_e + \left(-\frac{2}{3} - \frac{1}{3}\right)q_e = -q_e ,$$

so that charge is indeed conserved. This decay does not involve any electrons or neutrinos, so all lepton numbers are zero before and after, and the lepton numbers are unaffected by the decay.

(c) Using Table 31.4, we can write the equation in terms of its constituent quarks:

$$\boxed{(sss) \rightarrow (uds) + (\bar{u}s)} \text{ or } s \rightarrow u + \bar{u} + d .$$

Since there is a change in quark flavor, the weak nuclear force is responsible for the decay.

31.31 (a) From Table 31.4, we know the quark composition of each of the particles involved in the decay:

$$\Sigma^-(dds) \rightarrow n(udd) + \pi^-(\bar{u}d) .$$

The charge is conserved at -1. The baryon number is conserved at $B = 1$. All lepton numbers are conserved at zero, and finally the mass initially is larger than the final mass: $m_{\Sigma^-} > (m_n + m_{\pi^-})$, so, $\boxed{\text{yes}}$, this decay is possible by the conservation laws.

(b) Using Table 31.4, we can write the equation in terms of its constituent quarks:

$$\boxed{dds \rightarrow udd + \bar{u}d} \text{ or } s \rightarrow u + \bar{u} + d$$

31.37 (a) The energy released from the reaction is determined by the change in the rest mass energies.

$$Q = (mc^2)_i - \Sigma(mc^2)_f = (m_p - m_{\pi^0} - m_{e^+})c^2 .$$

Using Table 31.2, we can then determine this difference in rest mass energies:

$$Q = (938.3 \text{ MeV}/c^2 - 135.0 \text{ MeV}/c^2 - 0.511 \text{ MeV}/c^2)c^2 = \boxed{802.8 \text{ MeV}}$$

(b) The two gammas will carry a total energy of the rest mass energy of the π^0:

$$\pi^0 \rightarrow 2\gamma \Rightarrow Q_{\pi^0} = m_{\pi^0}c^2 = 135.0 \text{ MeV} .$$

The positron/electron annihilation will give off the rest mass energies of the positron and the electron:

$$e^- + e^+ \rightarrow 2\gamma \Rightarrow Q_{e^+} = 2m_e c^2 = 2(0.511 \text{ MeV}) = 1.022 \text{ MeV} .$$

So, the total energy would be the sum of all these energies:

$$Q_{tot} = Q + Q_{\pi^0} + Q_{e^+} = \boxed{938.8 \text{ MeV}}$$

(c) Because the total energy includes the annihilation energy of an extra electron. So the full reaction should be:

$$p + e \rightarrow (\pi^0 + e^+) + e \rightarrow 4\gamma .$$

31.43 (a) To determine the number of particles created, divide the cosmic ray particle energy by the average energy of each particle created:

$$\text{\# of particles created} = \frac{\text{cosmic ray energy}}{\text{energy/particle created}} = \frac{10^{10}\ \text{GeV}}{\left(0.200\ \text{GeV}/c^2\right)c^2} = \boxed{5 \times 10^{10}}\ .$$

(b) Divide the number of particles by the area they hit:

$$\text{particles/m}^2 = \frac{5 \times 10^{10}\ \text{particles}}{\left(1000\ \text{m}\right)^2} = \boxed{5 \times 10^4\ \text{particles/m}^2}\ .$$

31.49 (a) On average, one proton decays every 10^{31} years $= 12 \times 10^{31}$ months. So for one decay to occur every month, you would need:

$$N\left(\frac{1}{1.2 \times 10^{32}\ \text{months/decay}}\right) = \frac{1\ \text{decay}}{\text{month}} \Rightarrow N = 1.2 \times 10^{32}\ \text{protons}\ .$$

Since you detect only 50% of the actual decays, you need twice this number of protons to observe one decay per month, or $N = 2.4 \times 10^{32}$ protons. Now, we know that one H_2O molecule has 10 protons (1 from each hydrogen plus 8 from the oxygen), so we need 2.4×10^{31} H_2O molecules. Finally, since we know how many molecules we need, and we know the molar mass of water, we can determine the number of kilograms of water we need.

$$\left(2.4 \times 10^{31}\ \text{molecules}\right)\left(\frac{1\ \text{mole}}{6.02 \times 10^{23}\ \text{molecules}}\right)\left(\frac{0.018\ \text{kg}}{\text{mole}}\right) = \boxed{7.2 \times 10^5\ \text{kg of water}}$$

(b) Now, we know the density of water from Table 10.1, so we can determine the volume of water we need:

$$V = m\rho = \left(7.2 \times 10^5\ \text{kg}\right)\left(\frac{1\ \text{m}^3}{1000\ \text{kg}}\right) = \boxed{7.2 \times 10^2\ \text{m}^3}$$

(c) If we had 7.2×10^2 m^3 of water, and the actual decay rate was 10^{33} y, rather than 10^{31} y, a decay would occur 100 times less often, and we would have to wait on average $\boxed{100\ \text{months}}$ to see a decay.

FRONTIERS IN PHYSICS

(32)

CONCEPTUAL QUESTIONS

32.1 If you are a point on a balloon that is expanding, it looks as if everyone is moving away from you. That does not mean that you are the center of expansion, because similarly any other point on the balloon would see the same behavior. The fact that the universe is expanding means everyone everywhere feels that they are the center of the expansion of the universe.

32.4 If the red shift were due to gravitational fields, the red shift would be proportional to the size of the galaxy (because the size is proportional to the mass). This gravitational effect is also independent of distance. Thus, the galactic red shifts do not arise from high gravitational fields.

32.7 The wrinkles in the CMBR should have originally been caused by quantum fluctuations. Stronger gravitational forces caused by this initial concentration would reinforce the regions of dense energy. The fluctuations are real, but smaller than originally expected.

32.10 The farther away a star is, the less of the light emitted would reach our telescope. Therefore, objects farther away appear less bright than closer objects that have the same intensity. For far away galaxies, the wavelength with greatest intensity would be red-shifted. We must make sure to look in the infrared spectrum to make an accurate comparison. Since we are not sure of exact distances to these galaxies, our initial calibrations may not be accurate.

32.13 The images are of the same object because they all have the same spectrum. The outer images will have a larger red shift because light lost energy as it bent in the gravitational well of the intermediary galaxy. If we correct for this effect, all images from the same object will have the same red shift, even though they have traveled different paths. Hence, the distance traveled through space does not cause cosmological red shifts.

32.16 The lifetime of a black hole is many, many orders of magnitude longer than the life of the universe (5 billion years). Since large black holes are relatively young, we do not expect to see such radiation. For small black holes, the predicted radiation is very small and not easily observable.

32.19 If neutrinos are massless they travel at the speed of light and time will not pass for them, so that they cannot change without an interaction.

32.22 Complex adaptive systems are more interesting; compare the evolution of life versus the orbit of a complex pendulum. However, interesting is in the eye of the beholder.

32.25 Good thermal contact means that two objects are at the same temperature. Since liquid nitrogen can never get warmer than 77K, an object in good thermal contact with liquid nitrogen will also be no warmer than 77K.

32.28 Physicists, by their very nature and large egos, have always defined physics as that which can be explained tomorrow only by their shear genius. Therefore, no true physicist would admit not being able to completely understand the laws of physics, but likewise, they will never actually get there.

PROBLEMS

32.1 The approximate mass of the luminous matter in the Milky Way galaxy can be found by multiplying the number of stars times 1.5 times the mass of our sun (found in the front cover of the textbook):

$$M = \left(10^{11}\right)(1.5)m_s = \left(10^{11}\right)(1.5)\left(1.99 \times 10^{30} \text{ kg}\right) = \boxed{3 \times 10^{41} \text{ kg}}$$

32.7 (a) Using Equation 32.1, and the Hubble constant, we can calculate the approximate velocity of the near edge of the known universe:

$$v = H_c d = \left(20 \text{ km/s/Mly}\right)\left(10 \times 10^3 \text{ Mly}\right) = \boxed{2.0 \times 10^5 \text{ km/s}}$$

(b) To calculate the fraction of the speed of light, divide this velocity by the speed of light:

$$\frac{v}{c} = \frac{\left(2.0 \times 10^5 \text{ km/s}\right)\left(10^3 \text{ m/km}\right)}{3.00 \times 10^8 \text{ m/s}} = 0.67 \text{ , so that } \boxed{v = 0.67c}.$$

32.13 The relative brightness of a star is going to be proportional to the ratio of surface areas times the luminosity, so that:

$$\text{Relative Brightness} = \left(\text{luminosity}\right)\frac{4\pi r_{sun}^2}{4\pi R_{Andromeda}^2} = \left(10^{12}\right)\left(\frac{r_{Sun}}{R_{Andromeda}}\right)^2.$$

From the front cover of the textbook we know the average distance to the sun is $1.496 \times 10^{11} \text{ m}$, and we are told the average distance to Andromeda, so:

$$\text{Relative Brightness} = \frac{\left(10^{12}\right)\left(1.496 \times 10^{11} \text{ m}\right)^2}{\left[\left(2 \times 10^6 \text{ ly}\right)\left(9.46 \times 10^{15} \text{ m/ly}\right)\right]^2} = \boxed{6 \times 10^{-11}}.$$

Note: this is an overestimate since some of the light from Andromeda is blocked by its own gas and dust clouds.

32.19 A star orbiting its galaxy in a circular orbit feels the gravitational force acting toward the center, which is the centripetal force (keeping the star orbiting in a circle). So, from Equation 8.8, $F = G\dfrac{mM}{r^2}$, we get an expression for the gravitational force acting on the star, and from Equation 8.7, $F_c = m\dfrac{v^2}{r}$, we get an expression for the centripetal force keeping the star orbiting in a circle. Setting the two force expressions equal gives:

$$F = \frac{mv^2}{r} = \frac{GMm}{r^2},$$

where m is the mass of the star, M is the mass of the galaxy (assumed to be acting from the center of the rotation), G is the gravitational constant, v is the velocity of the star, and r is the orbital radius. Solving the equation for the velocity gives:

$$\boxed{v = \sqrt{\frac{GM}{r}}},$$

so that the velocity of a star orbiting its galaxy in a circular orbit is indeed inversely proportional to the square root of its orbital radius.

32.25 (a) Since the Hubble constant has units of km/s/Mly, we can calculate its value by considering the age of the universe and the average galactic separation. If the universe is 10^{10} years old, then it will take 10^{10} years for the galaxies to travel 1 Mly. Now, to determine the value for the Hubble constant, we just need to determine the average velocity of the galaxies from Equation 2.3: $v = \dfrac{d}{t}$, so that

$$v = \frac{1\ \text{Mly}}{10^{10}\ \text{y}} = \frac{1 \times 10^{6}\ \text{ly}}{10^{10}\ \text{y}} \times \frac{9.46 \times 10^{12}\ \text{km}}{\text{ly}} \times \frac{1\ \text{y}}{3.156 \times 10^{7}\ \text{s}} = 30\ \text{km/s} .$$

Thus, the Hubble constant is approximately:

$$H_{c} = \boxed{\frac{30\ \text{km/s}}{1\ \text{Mly}}} .$$

(b) Now, the time is 2×10^{10} years, so the velocity becomes:

$$v = \frac{1\ \text{Mly}}{2 \times 10^{10}\ \text{y}} = \frac{1 \times 10^{6}\ \text{ly}}{2 \times 10^{10}\ \text{y}} \times \frac{9.46 \times 10^{12}\ \text{km}}{\text{ly}} \times \frac{1\ \text{y}}{3.156 \times 10^{7}\ \text{s}} = 15\ \text{km/s} .$$

Thus, the Hubble constant would be approximately:

$$H_{c} = \boxed{\frac{15\ \text{km/s}}{1\ \text{Mly}}} .$$